OUR HOUSE
IN THE CITY

NEW URBAN
HOMES AND
ARCHITECTURE

gestalten

AN URBAN DWELLING REVIVAL

■   Here it is—your dream location. The address can't be beat. It's just a short stroll from the city center and close to the best bars, restaurants, supermarkets, a school for the kids, everything you were hoping for. Now here's the trouble. Even though you couldn't imagine a better location, you could imagine a better site. This site in particular is a narrow infill plot, just 20 feet wide. The towering neighbors on either side also seem to block out most of the light, which was always an important criterion for you. This optimal location also means that privacy may be an issue. With both the nearby neighbors close at hand and a steady stream of pedestrians passing by, it could become difficult to ever really feel at home. But then again, you don't find a dream location every day. So what do you do?

These contemporary townhouses not only masterfully transform site challenges into assets but also inspire a more efficient and streamlined lifestyle for their occupants.

As the classically perfect site proves few and far between, solving the urban puzzle that remains stands at the heart of determining the success of a city's residential revival. After the rise of suburbia and the decline of our downtowns as vibrant social centers, the city is once again back in vogue. Choosing character and location over sprawling building footprints, many homeowners now prefer trading in their McMansion in the outskirts for their own personal slice of the great and bustling metropolis. This urban renaissance opens the door for an ingenious new breed of city dwellings—the 21st century townhouse. These contemporary townhouses not only masterfully transform site challenges into assets but also inspire a more efficient and streamlined lifestyle for their occupants.

Where constraints exist, novel solutions never stray too far behind. **OUR HOUSE IN THE CITY** provides an in-depth look into the stories, dreams, and strategies developed for a range of residential projects that embrace their place within the city. These inventive and culturally sensitive houses appear in urban locations around the world as the result of extensive collaborations between client and architect. Whether digging into a residential hillside, proudly commanding a corner lot, or nestling between pronounced neighboring structures, each location represents a new opportunity to find one's place within the given context.

The modern day townhouse views context as a creative challenge rather than a limitation. By finding new ways to bring light, nature, and air into the building, the rethinking of previously untenable sites resurrects swatches of unused lands across the city and reinstates them back into the urban fabric. The Japanese remain at the forefront of the current guard of architects re-imagining the townhouse. Masters of the micro, the Japanese not only excel in doing more with less but also find countless ways of making small spaces feel big.

MON FACTORY/HOUSE by Eastern Design Office choreographs an iconic live/work residence that grants a traditional crestmaker rare access to both nature and sky. With the store placed on the ground level, the workshop and living spaces rise up to the second floor. Two voids carve out of this top level to form generous courtyards. These multipurpose outdoor spaces link the workshop to the open kitchen, living, and dining room, and this social area to a private bedroom in the back. Flooding the interior with daylight and natural ventilation, these hidden courtyards support a meaningful and tranquil connection to the outdoors in one of the densest cities in the world. The serene **OPTICAL GLASS HOUSE** by Hiroshi Nakamura grows out of one Hiroshima's busiest thoroughfares. Surrounded by tall buildings on both sides, the house utilizes a mesmerizing façade composed of 6,000 glass blocks that buffer the sights and sounds of the street life just outside. The idiosyncrasies of the transparent bricks enable the family to quietly observe the happenings of the city from inside their ethereal refuge without being exposed to the public eye. From the street, this same glass wall presents only an abstracted image of the tall trees swaying in the courtyard within. Such an elegant cosmopolitan garden solution allows its urban residents the unusual luxury of cultivating a life in the city where all aspects of the home interact with nature.

A delicate balance between standing out and blending in serves as a key parameter guiding the design of the revamped townhouse. Such clever and visionary solutions strike a satisfying balance between reflecting the individuality of the client and connecting with the local vernacular and rhythm of the city. Black Line One X Architecture Studio's **PROFILE HOUSE** in Melbourne illustrates the scope of interpretative freedom that comes from working with both contextual cues and the desires of the home owner. In this instance, the two-faced renovation and extension dialogues with both sides of its corner location. Retaining its traditional craftsman style appearance from the front of the site, the addition introduces a multipurpose contemporary silhouette along the side. Evolving from a fence for the old house to a wall of the new addition, to a door for the back garden, the angular lines shield the private residential oasis just behind. At the same time, this very intimate enclosure still retains a very public face that purposefully references the familiar industrial outlines found in this transitioning warehouse district. **HOUSE 77** by dIONISO LAB takes a narrow infill plot on a densely packed residential street and turns it into a compelling cultural homage to the historical Portuguese fishing town. The slender, three-story home applies a graphic folding screen over the street façade. Its metallic surface mitigates the levels of privacy desired by the family within while resurrecting the primitive symbols of the region. Traditionally used as a way to communicate between fishermen, these timeless symbols cut out of the screens not only pay tribute to the rich heritage of the neighborhood but also create a nuanced and one-of-a-kind light show from inside the house. Flawlessly bridging the gaps between past, present, and future, these two residences devise innovative ways of relating to one's city and to one's home.

By designing for maximum utility, many projects completely upend conventional expectations for what a house can and should look like. With some doing away with windows entirely in favor of sophisticated skylight systems, others find ways of integrating nature and the outdoors into all aspects of the interior. The Spanish office Grupo Aranea's experiential rooftop residence, **CASA LUDE**, strips away conventions. Reading as a type of futuristic UFO precariously perched above the original family home, this new add-on substitutes a network of rooftop terraces for the classical window. Now, every aperture relates to a usable exterior space, promoting not only natural light and air circulation but also encouraging an active relationship to the outdoors and surrounding city.

New models for the single-family, small lot house capitalize on their spatial constraints. The diversification of modest building footprints leads to myriad space-saving options. Whether integrating flexible partitions or smartly stacking spaces and programs, the importance of multi-use spaces plays a crucial role in a home's ability to respond to the changing needs of a family over time. Mark Koehler Architect's **HOUSE LIKE VILLAGE** in Amsterdam addresses the intricate rituals of daily life. Comprised of a network

By designing for maximum utility, many projects completely upend conventional expectations for what a house can and should look like.

of multipurpose corridors with shared public mezzanines above, the malleable layout easily adjusts to the family's organic evolution. Just as the number of occupants in a house expands or contracts, the nature of their activities continue to mature and intertwine. The act of designing with this inherent flux

of townhouses that deal with a variety of plots and their site-specific challenges. Showcasing solutions for the narrow infill and corner lots, the book also covers everything from the latest strategies in hillside developments to the more suburban residential endeavors and experimental high-density rooftop add-ons. The scope of the projects extends from entirely new constructions to thoughtful renovations and additions. Each residence, whether visually commanding or all but invisible, designs for longevity, adaptability, and the shifting needs of its clients.

Many different ways exist for making a standard or limited space your own. The art of personalizing within an established context manifests itself in countless forms.

in mind minimizes the superfluous floor plan by tapping into and working around the real concerns and interests of the client.

In addition to exalting a more humble scale for living, the new townhouse also embeds a spectrum of sustainable features. These sustainable elements can include more simple concepts such as natural ventilation, efficient insulation, recycled materials, and sunscreens for minimizing solar gain to more complex systems for generating solar powered heating and cooling. Versatile and economical solutions for improving general efficiency around the house can be found for any and all budgets. Such practical alternatives go a long way toward not only improving architecture's relationship to the environment but also providing lasting, long-term energy saving

The 21st-century townhouse captures the allure of the city. When considering the value of our urban centers the great thinker Jane Jacobs comes to mind. In her seminal book **THE DEATH AND LIFE OF GREAT AMERICAN CITIES,** Jacobs describes the power of the metropolitan community as she writes, "Cities have the capability of providing something for everybody, only because, and only when, they are created by everybody." This illuminating statement emphasizes the unique opportunity that downtown living affords the urban dweller—a chance to shape and be shaped by the pulsing heartbeat of the city. Let's take comfort in the limitations, the unexpected, and the unpredictable and transform them into a spectacular way of living. You may never have the perfect site, so how do you make the site perfect for you?　■

Each residence, whether visually commanding or all but invisible, designs for longevity, adaptability, and the shifting needs of its clients.

results. With many of the most playful projects being the most sustainably minded, the era of equating energy efficiency with unaesthetic solutions is quickly becoming a notion of the past.

Many different ways exist for making a standard or limited space your own. The art of personalizing within an established context manifests in countless forms. When carving one's own nook out of the metropolis, each distinct site suggests its own bespoke solution. **OUR HOUSE IN THE CITY** showcases a range

CONTENT

INTRODUCTION
003 – 005

006 – 007

Studio Velocity
MONTBLANC HOUSE
008 – 013

Spacecutter
CARVED DUPLEX PROJECT
014 – 019

Andrew Maynard Architects
HOUSE HOUSE
020 – 027

Eastern Design Office
MON FACTORY/HOUSE
028 – 035

Takeshi Hosaka Architects
DAYLIGHT HOUSE
036 – 041

Hiroshi Nakamura & NAP
OPTICAL GLASS HOUSE
042 – 049

TERRA E TUMA
MARACANÃ HOUSE
050 – 055

A21 Studio
THE NEST
056 – 063

Case-Real
WHITE DORMITORY

FOR IL VENTO
064 – 069

Edwards Moore
DOLLS HOUSE
070 – 075

Level Architects
SKATE PARK HOUSE
076 – 083

Moonhoon
LOLLIPOP HOUSE
084 – 089

Andrew Maynard Architects
VADER HOUSE
090 – 095

Black Line One X
Architecture Studio
PROFILE HOUSE
096 – 103

Tato Architects
HOUSE IN YAMASAKI
104 – 109

Fran Silvestre Arquitectos
HOUSE ON MOUNTAIN-SIDE OVERLOOKED BY CASTLE
110 – 117

SNARK
HOUSE IN KEYAKI
118 – 123

Areal Architecten
HOUSE DV WILRIJK
124 – 129

Lischer Partner Architekten
STADTVILLEN
130 – 135

Meixner Schlüter Wendt
Architekten
RESIDENCE Z
136 – 141

Jonathan Tuckey Design
SHADOW HOUSE
142 – 147

Brandt + Simon Architekten
SCHUPPEN
148 – 153

Braun und Güth Architekten
PÜNKTCHEN
154 – 159

C. Fischer Innenarchitekten
TOWNHOUSE
160 – 167

bfs d
GAS STATION
168 – 173

Hideaki Takayanagi
LIFE IN SPIRAL
174 – 179

Marc Koehler Architects
HOUSE LIKE VILLAGE
180 – 187

Alphaville
NEW KYOTO TOWN HOUSE
188 – 193

J. MAYER H.
OLS HOUSE
194 – 199

mA-style architects
LIGHT WALLS HOUSE
200 – 205

Takeshi Hosaka
OUTSIDE IN
206 – 211

dIONISO LAB
HOUSE 77
212 – 219

Hiroshi Nakamura & NAP
HOUSE SH
220 – 227

Studio Velocity
HOUSE IN CHIHARADA
228 – 235

Kimihiko Okada
TODA HOUSE
236 – 241

Clare Cousins Architects
BRICK HOUSE
242 – 247

Grupo Aranea
CASA LUDE
248 – 253

INDEX
254 – 255

IMPRINT
256

MONTBLANC HOUSE

Okazaki, Japan, 2009

An iconic white residential mountain rises out of a Japanese neighborhood. This slanted structure acts as both a home and a workplace, with a small beauty parlor situated on the ground floor. Designed for a family of four, the young couple runs the shop below and lives in the spaces above together with their three-year-old daughter and newborn baby. Set back from the street to accommodate parking for the beauty parlor in front, the house stands as an undeniably elegant example of a live/work space.

The location, surrounded by neighboring houses in three directions, is situated in a quiet residential area. While most houses in the vicinity only have two stories, with views of the mountains in the distance, this configuration proves inappropriate given the constraints of this particular site. As the neighboring houses and apartments stand close to the site, their windows open on to it, leaving little privacy for the new family. This initial lack of privacy represents the main consideration influencing the final design of the home. The surprising outcome succeeds in creating a private yet open space inside such a tightly enclosed site.

A continuous exterior space extends from the first to the third floor under a large inclined roof. The building's gable shape includes five generous openings in

the angled roof that bring sufficient natural light, air, and scenery into the building. Lacking sashes, the windows act as dramatic voids that take on a completely different scale from the windows of the surrounding houses. This unique approach to privacy and access results in a wondrously surreal sense of scale—a dollhouse enlarged.

The one-of-a-kind home captures the intimate qualities present in a walk through a forest or a stroll through a dense thicket. **A NATURAL REFUGE IN THE HEART OF THE CITY, THE PROJECT GENTLY EMBRACES**

ITS RESIDENTS IN A PROTECTIVE SHELL. This light barrier stays in tune with the nuances of the natural world, framing swatches of blue sky as it streams through the leaves. These ethereal interpretations of indoor/outdoor spaces allows the family to hear

the hushed murmur of the wind as it passes through the trees and the sound of cheerful birds calling out to their mates. Such a rare experience of nature within the city makes this house a truly exceptional example of contextual sensitivity.

Beneath the giant A-frame, one encounters the outside at various points throughout the interior. This expansive, luminous roof introduces a number of

covered, semi-exterior spaces as the residents ascend through the house. The tilted planes of the roof enable it to become both ceiling and wall. These sloping walls curate views up towards the sky. Doing away with the traditional, horizontally aligned window, these portals to the outside world transport the family into a serene and peaceful setting that gradually filters out the bustling city on all sides. This clever window treatment carves out a moment of respite without needing to set foot outside the city. Tranquil and rejuvenating, the home's formal approach transcends expectations and defies the conventions for an urban dwelling.

The layout establishes a diversity of exteriors that correspond to the various angles and heights throughout the site. Manifesting as windows, terraces, and balconies, these transcendent experiences of the outdoors gradually block out all views from the neighboring plots as well as the memory of the neighboring houses' existence. This precious feeling of privacy and connection to nature serves as a timely reminder that quality of life need not be sacrificed in order to have a home in the city.

EFFICIENTLY ACTIVATING EVERY SURFACE, THE DAZZLING ALL-WHITE INTERIOR CONSISTS OF A CORE OF TRUE INTERIOR SPACES STRADDLED BY A SERIES OF INDOOR/OUTDOOR TERRACES. These terraces permeate the left side of the house in a network of lofted mezzanines. Each level gradually heightens its relationship to nature. Beginning on the ground floor, this first, narrow but lofty terrace reads as an extension of the garden just outside the perimeter walls. This more private garden brings the gravel and oversized, pebble-shaped stepping stones found in the driveway into this transitional area. The secret garden in fact doubles as the private entry to the house, with the shop entrance just a few steps away. One level up, the single terrace splits into two unique outdoor spaces that straddle the main children's room. The visionary placement of these terraces rethinks the idea of the classic backyard, presenting an exhilarating alternative for the more compact city townhouse. Culminating in a resplendent roof terrace, this shaded loft area debunks yet another residential typology—the attic. Typically the most dark and unappealing part of the house, this attic space instead transforms into the residence's most compelling feature. Dark wooden furniture gracefully contrasts with the brilliant surroundings, while trees in silver pots peek out through the numerous square cutouts on the pitched roof. Bordered by pristine white walls on every side, these sizable windows to the world behave as shifting canvases painted with sky. ∎

PLANS & DRAWINGS

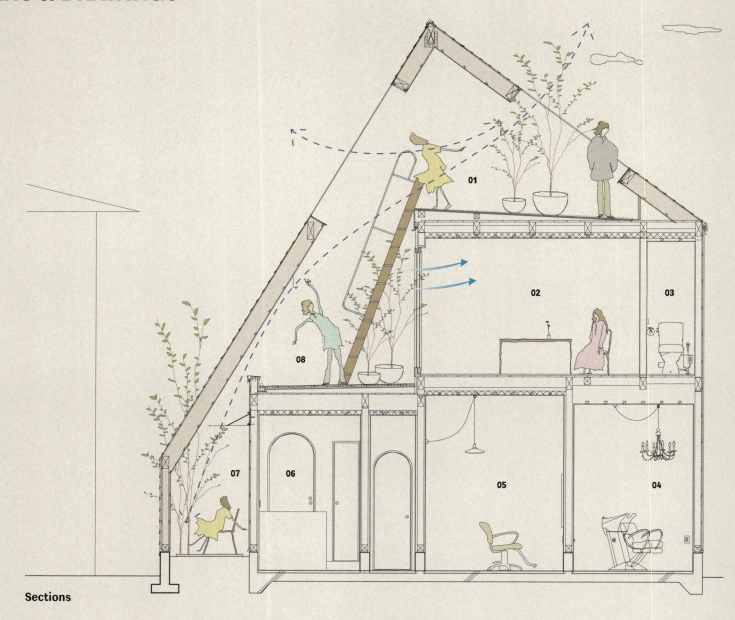

Sections

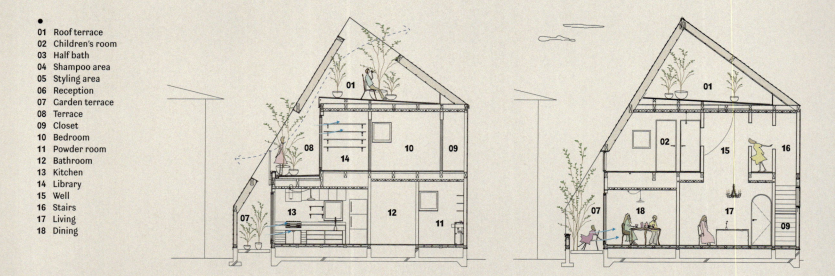

●
01 Roof terrace
02 Children's room
03 Half bath
04 Shampoo area
05 Styling area
06 Reception
07 Garden terrace
08 Terrace
09 Closet
10 Bedroom
11 Powder room
12 Bathroom
13 Kitchen
14 Library
15 Well
16 Stairs
17 Living
18 Dining

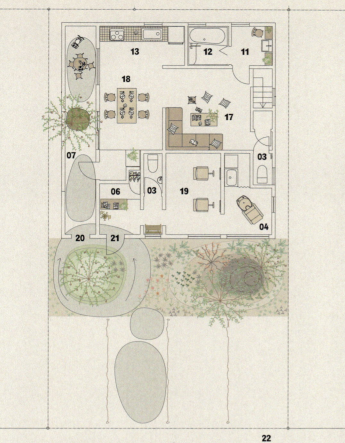

Ground floor plan

Second floor plan

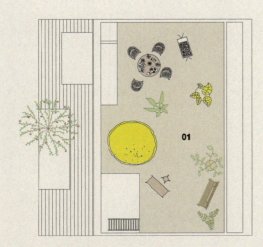

Third floor plan

MONTBLANC HOUSE

Spatial Strategy

A slanted roof brings nature into all three levels of the house. Growing in size from bottom to top, these indoor outdoor spaces bring in light and air while maintaining privacy between the residence, the on-site hair salon, and the neighborhood.

●

Number of square meters:
123.44
Number of rooms:
6 rooms, 1 bathroom
Number of floors:
3
Number of residents:
2 adults, 2 children
Sustainable features:
Natural ventilation
Outdoor areas/rooftop access:
Indoor outdoor courtyards on every level
Context:
Residential area
Type:
New construction

●

01 Roof terrace
02 Children's room
03 Half bath
04 Shampoo area
05 Styling area
06 Reception
07 Garden terrace
08 Terrace
09 Closet
10 Bedroom
11 Powder room
12 Bathroom
13 Kitchen
14 Library
15 Lightwell
16 Stairs
17 Living
18 Dining
19 Shop
20 Approach to house
21 Approach to shop
22 Street

Spacecutter
CARVED DUPLEX PROJECT
Brooklyn, USA, 2013

TO LARGE MONOLITHS IN VARIOUS MATERI-AL FORMS. Built within a classic New York townhouse shell, the extensive renovation slowly unfolds from a black bedroom on the lower level to a white living room on the upper level. The process of working within an older tenement structure introduces significant structural challenges. Requiring a range of reinforcements for the addition, the urban dwelling features updated floors, foundations, and masonry parapet. Meticulously restored, both roof and the original cornice now shine with a second life.

Built for a husband and wife, the eight-room, two-level home mixes rich materials with black, grey, and white massive forms. One example of such contrast occurs in the alluring bathroom where handmade, sapphire blue tiles glimmer like jewels against the white tiles and a recess carved for storage. Brimming with detail and creative transitions from one space to another, grayish wood floors wrap up and over walls to direct the couple upstairs or across hallways.

Four square windows illuminate an open kitchen and dining area. Here, two prominent triangular-shaped volumes spill into the room. The first triangle holds the stairs that connect to the addition above. A second and more sculptural wedge integrates the stairs that link the entrance to the duplex with the public hall. Together, these bold forms generate a striking entry sequence that resolves upon reaching the

■ 　Carved Duplex comprises a rooftop addition to a brick townhouse in Brooklyn's Williamsburg neighborhood. The one-bedroom apartment takes over the thirdrd floor of the building and expands it upwards one level via a Cor-Ten steel box. High impact with a modest footprint, the layout combines two railroad apartments to form the lower level. The compact but intricate project includes two twin-cut fireplaces, a large, wedge-shaped core, a stainless steel kitchen, a long, stainless steel tub, grey wooden floors, and a coveted 28-square-meter roof deck.

THIS RESIDENTIAL PROJECT EXPLORES THE PHENOME-NOLOGICAL QUALITIES OF LIVING IN CLOSE PROXIMITY

more open dining area. An expansive, custom table design by the architect dominates this dining space. The wooden table, referred to as the Monolith Table, boasts slender red cedar planks that elegantly erode to tuck in four chairs, two benches, and two booths. Exposed brick on either side showcases the original fireplaces uncovered during demolition.

Echoing the massing of the Monolith Table, a minimalist kitchen eliminates the need for overhead cabinets. Instead, all storage is consolidated in the stainless steel floating island and the wood clad cube that also houses the refrigerator, pantry, and third-floor bathroom. Backed by silver travertine, a carved recess forms a seating nook above the custom stainless steel bathtub for relaxation.

A brief wooden hallway transitions from the bathroom area into the master bedroom and dressing room. Shifting the quality of light, the gray wood walls fade to black in order to encourage sleep. Inside this darker space, a bookshelf cut from the drywall reveals the exposed brick just behind. This bedroom serves as the home's most secluded space. Nestled behind the massive stairs, the triangular core, and the bath cube, the master bedroom stays shielded from the street—the ultimate quiet retreat.

THE ICONIC EXTERIOR MASS OF THE FOURTH FLOOR, MADE FROM CLASSIC COR-TEN STEEL, GAINS ITS DISTINCT CHARACTER AS IT SLOWLY RUSTS OVER TIME. Functioning as a bright multipurpose living space, this square box gradually reveals its golden patina as it ages, referencing the warm tones of the original building's brickwork. The first of its kind, the updated traditional façade stands as the first ventilated Cor-Ten steel rainscreen envelope in New York City. Inside these reddish walls, a modern living room and entertainment area coexist with the large outdoor roof deck. The streamlined interior easily converts into an additional bedroom suite depending on the changing needs of the couple.

The choice to renovate and add onto the more than 150 year old structure leads to the reuse and salvaging of many original units. This new spatial strategy highlights the extant brick walls and fireplaces that date back to the building's inception in the 1860's.

By repurposing the existing structure, the project's carbon footprint decreases considerably—adding up to savings of approximately 35 tons of carbon dioxide (CO_2). The significant amount of metal used throughout the residence contains recycled content and can be recycled again if ever dismantled. These recyclable elements include the Cor-Ten steel exterior, stainless steel kitchen cabinets and bathtub, steel joists, framing, columns, and beams, and finally an aluminum exterior deck with Cor-Ten steel guardrails. This inviting roof deck is situated off the upstairs living room and receives direct sunlight from noon to sunset. From this private lookout, the couple can use their outdoor grill or simply share a rare moment of respite over the rooftops of the pulsing metropolis. ∎

PLANS & DRAWINGS

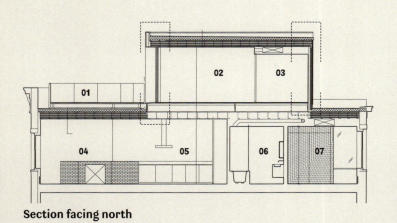

Section facing north

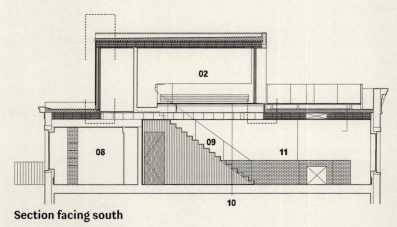

Section facing south

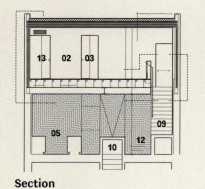

Section

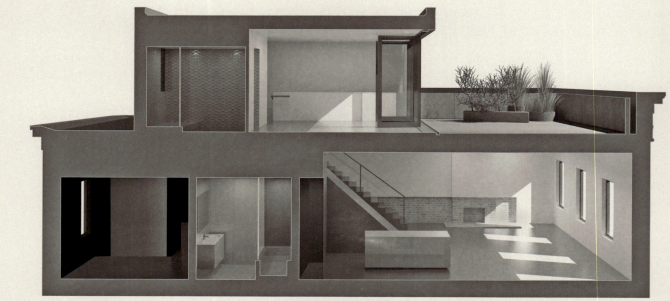

Carved duplex

●
01 Roof deck
02 Family room
03 Study
04 Dining room
05 Kitchen
06 Bathroom 1
07 Changing room
08 Bedroom
09 Stairs 2
10 Stairs 1
11 Living room
12 Entry hall
13 Bathroom 2

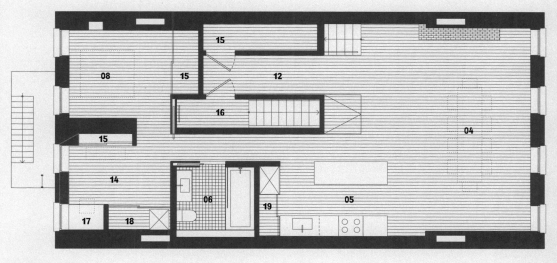

First floor plan

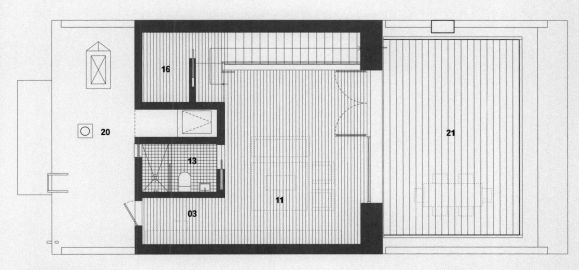

Second floor plan

CARVED DUPLEX PROJECT

Spatial Strategy

An addition of a steel cube to the top of an old brownstone expands the interior and introduces a rooftop deck. This layout slowly unfolds from a black bedroom on the lower level to a white living room on the upper level.

●

Number of square meters:
214
Number of rooms:
8
Number of floors:
2
Number of residents:
2
Sustainable features:
Repurposed original structure, recycled building materials
Outdoor areas/rooftop access:
Roof deck
Context:
City center
Type:
Addition and renovation

●

01 Roof deck	**14** Dressing room	
02 Family room	**15** Closet	
03 Study	**16** Public hall	
04 Dining room	**17** Make-up vanity	
05 Kitchen	**18** Laundry/closet	
06 Bathroom 1	**19** Pantry	
07 Changing room	**20** Roof/mechanical area	
08 Bedroom	**21** Outdoor deck	
09 Stairs 2		
10 Stairs 1		
11 Living room		
12 Entry hall		
13 Bathroom 2		

Andrew Maynard Architects
HOUSE HOUSE
Richmond, Australia, 2012

These neighboring terrace homes are owned by two generations of one family. As both houses were in need of repair and update, the project consists of a single building that extends both homes while still relating them back to each other. This visionary redesign on a corner lot produces two separate homes within one architecture. The striking new golden wood addition runs north/south while the original houses run east/west. Inspiring communal living, a glass partition between each terrace slides away to reveal one large backyard.

THIS RESIDENTIAL PROJECT TAKES A STAND AGAINST THE AUSTRALIAN TENDENCY TOWARDS LARGE, FLAT, AND LOW DENSITY HOUSING. With few topographical constraints to force homes to have a small footprint, the suburban typology becomes one of increased isolation and car dependency. Architect Andrew Maynard elaborates on this housing epidemic as he explains:

"In Australia we go wide and low. We pancake our homes. We eat up our outdoor space. Often people move to the suburbs under the false logic that they will have an abundance of open space and room for kids to play; however, the enormous size of houses now makes this a convenient myth rather than a true outcome."

This statement underlines the deliberate decision to impose a vertical logic onto these houses.

By introducing a footprint restriction far beyond the requirements, the tall, thin arrangement maximizes what would otherwise be a modest backyard on a small site.

Instead of crashing new structures into old, the design deliberately creates two separate forms with a clear gap in between. Glass infills keep weather out of this cavernous space. Here, a spiral staircase rises up between the two structures, interacting with both the aged brick of the terrace and the red cedar cladding of the new addition. Avoiding the use of new synthetic, shiny, or plastic materials, the elements

spaces, materials, and functions afford the homes the flexibility to adapt to the complex moods of their occupants. Stacked three levels high, the cavernous spaces with light cascading from above become a novel experience and refreshing alternative to the Australian residential typology. The active family living areas keep to an intimate size by integrating loose boundaries. This feeling of openness and interconnectedness grants the different spaces a light and atmospheric quality as they transition from past to present. The original front sitting room remains, followed by the shared living areas that transition into the dinning room and backyard. Cornices, rounded archways, Victorian ash cladding, and refined brickwork soften minimalist details and create a hybrid of old world and new world environments.

Both side fences can open up to let outdoor activity spill beyond the living area. This continuous relationship between indoor and outdoor living also manifests in the kitchen counter that extends through the rear glass wall. Ideal for large family gatherings, a built-in barbecue sits on the end of this counter. Grass and bricks intersperse to form a dynamic walkway that runs between the backyards of both houses. Just above, differently sized, square side windows haphazardly scattered over this back façade direct light into the various interior spaces of the house.

featured in this project exude a rich history. The new form, clad entirely in cedar, works in concert with the raw steel plate and detailing that describes the openings and thresholds between old and new. Dark plywood paneling rises through the light-filled void between the structures. The strategic use of mirrors on the cabinetry in the dining area makes the space feel large while giving the illusion that light is coming in from both sides. These mirrored surfaces also obscure and expand perceived boundaries, resulting in a mirage of continuous garden. The new structure is built across the rear of the Victorian terraces, which were respectfully repaired and reorganized.

THE KEY TO MAKING THESE MODESTT HOMES FLOURISH RELIES ON THE PROVISION OF A NUMBER OF SPACES WITH VARIOUS PERSONALITIES. These diverse

The levels above the living areas provide quiet contemplative spaces. Each room connects with both the rear yard and the internal light well. White walls and plank flooring lend the interior a traditional feeling while oscillating between minimalism and classicism. These spaces also explore themes of compression and expansion as they contract and open up to one another. Perforated mesh landings for each home's main spiral staircase generate a textured, multidimensional mood within the small spaces, turning everyone who walks by into an abstract painting.

Sustainability represents a reoccurring theme driving all of Andrew Maynard Architect's buildings and this project in particular. A core responsibility rather than a narrative tool, the sustainable features include double-glazed windows and a high-performance glass roof. Automated louvers stop sunlight

from directly hitting the glass to avoid the greenhouse effect typically associated with glass roofs. Ranging from full daylight to complete black out, the owners can adjust the louvers at anytime according to their preference. Louvers to the south of the light well are also automated to allow the space to quickly vent should heat build up. The new walls and solar panel covered roof are fitted with high-performance insulation, while the insulation of the existing terrace roofs was also upgraded.

A black graphic appears on the narrow cedar boundary wall as an homage to Melbourne's street art scene. Partially designed to deter graffiti, the graphic depicts a child like image of a suburban home. The playful overlapping of the classic family house silhouette on top of a more contemporary small lot development references the cozy interior mood just inside. A large lookout window prominently balances out the top of the composition of this slender façade. This prominent window transforms a bedroom into a tranquil lookout point to observe the happenings around the neighborhood. Capture an old-fashioned atmosphere within a new structure, the narrow footprint of this project maximizes intricacy without sacrificing imagination. ∎

PLANS & DRAWINGS

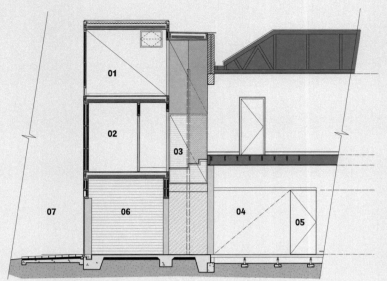

Section 1

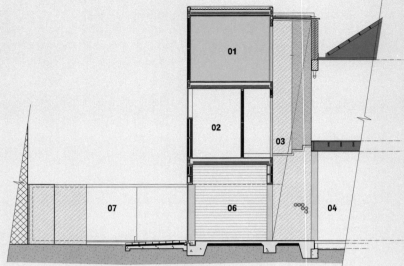

Section 2

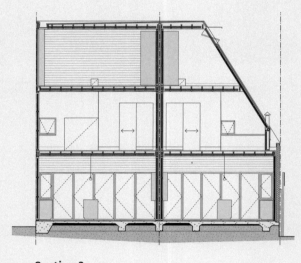

Section 3

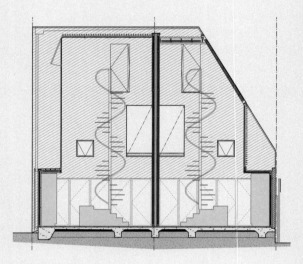

Section 4

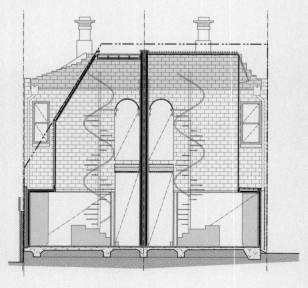

Section 5

●
01 Bedroom
02 Bathroom
03 Landing
04 Corridor
05 Half bath
06 Kitchen
07 Backyard

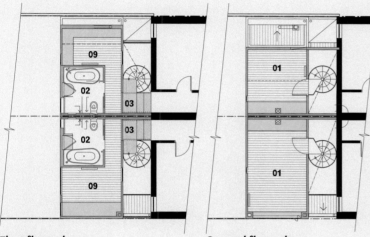

Ground floor plan

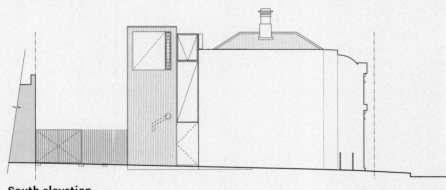

First floor plan   **Second floor plan**

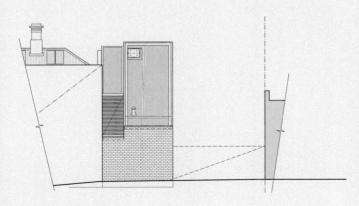

South elevation

Andrew Maynard Architects
HOUSE HOUSE

Spatial Strategy

The intentionally vertical design stacks spaces three levels high. Maximizing the backyard space on the small site, the house intentionally restricts its footprint, producing tall cavernous spaces that overflow with light.

●

Number of square meters:
180
Number of rooms:
8
Number of floors:
3
Number of residents:
2
Sustainable features:
Double-glazed windows, automated louvers for shading and climate control, high-performance insulation, rooftop solar panels
Outdoor areas/rooftop access:
Backyard
Context:
Inner suburb
Type:
Addition and renovation

North elevation   **West elevation**

●
01 Bedroom
02 Bathroom
03 Landing
04 Corridor
06 Kitchen
07 Backyard
08 Dining room
09 Study

Eastern Design Office
MON FACTORY/ HOUSE
Kyoto, Japan, 2007

Part monolithic storefront and part ethereal private residence, this single family, mixed-use home engages dramatic geometries and proportions as its driving aesthetic. The proud structure, built for a traditional craftsman and his daughter, functions as an abstract tribute to the client's exacting profession of applying crests onto classical Japanese clothing. Set on a narrow plot in a dense residential neighborhood of Kyoto, the home strikes an effective balance between connecting to the city and preserving the client's privacy and the introverted nature of his work. Accented by a warm wooden entry door, the concrete planes of the front façade overlap to symbolize the breast of a Kimono. This setback entry area introduces several panes of glass to add a level of transparency to the shopfront. The interior of this overlapping space holds the crestmaker's quiet yet busy shop, with the workspace and the living quarters nestled on the floor above. Within this entry area, the circular windows begin to double up. This unusual alignment of the portal windows across space creates theatrical lighting and enticing, three-dimensional lookout points for highlighted views of the city beyond. Proud yet secretive, the live/work structure reads as a mysterious bunker that invites closer inspection.

Twenty-six large, circular apertures run up and around the concrete exterior in a cross formation, rhythmically punctuating the massive volume to create a unique crest on the building. These round portal windows relate back to the work of the crestmaker, as every intricate crest pattern derives from a circle. As if applied by a stamp, two of these twenty-six circles are cut off at the top and side respectively. These truncated shapes correspond to the parts of the

façade that overhang or are offset from the interior space. The appearance of these two unfinished geometries generates a sequential evolution between the

graphic and functional elements of the exterior pattern. Allowing light to penetrate and drift though the interior and main stairs, the remaining twenty-four openings choreograph graceful, cheerful, and evolving interplays of light and shadow. Lending the project an aquatic theme, certain portal windows pull open like submarine doors to let air currents flow through the different spaces and levels of the house.

THE RESIDENTIAL AREAS OF THE HOME MOVE UP A LEVEL TO MAKE WAY FOR THE CREST SHOP IN THE FRONT AND A COVERED PARKING GARAGE IN THE BACK. In addition to bringing the house closer to the sky, the layout of the upper floor intersperses two outside courtyards in between its three main interior blocks. These courtyards channel wind, light, nature, and sky into the main living spaces.

From hanging plants to lines of shoes placed by the door, personal touches found throughout the house mitigate the muscular nature of the design. Beginning with the humble workroom at the front of the house, the layout then fluctuates between outdoor and indoor rooms. Located just off the landing next to the main stairs, this first modest workroom features a simple built-in table running along the back wall. Lit by a desk lamp and two portal windows, one above and one below the table, the cozy space inspires disciplined and focused work. A cheerful, glowing hedgehog floor lamp, a house plant, and a few framed artworks soften the austerity of the concrete surfaces.

In the midst of the heavy weight of the concrete, the house suddenly lightens to satisfy the craving for outdoor and open spaces. The strategic placement of courtyards on this upper residential level address the home's enhanced desire for the presence of organic and living highlights. **THIS VISCERAL TENSION BETWEEN NATURE AND THE MANMADE PRODUCES A CYCLICAL EXPERIENCE OF DESIRE THAT EBBS AND FLOWS AS ONE MOVES AROUND THE HOUSE.** The first courtyard connects the workroom to the kitchen and living spaces, acting as a place to relax and watch clouds drift by at the end of the day. Straddled by courtyards on either side, the multipurpose kitchen, dining, and living room offsets its rough concrete panels with floor to ceiling glass walls. These glass partitions open up the space to the outdoors, and behave as light wells to illuminate the dark interior

surfaces. The minimalist kitchen and living areas, comprised of light ash furniture, catch the glow of the passing sun. Two rows of built-in shelving, one slender and one thick, extend across the main wall. The top shelf spans the entirety of this wall. The contents of this shelf gradually transition from kitchen serving tools to more decorative design elements as it reaches the two sets of dining tables near the first courtyard. A grey sofa at the back of this family area demurely blends into the concrete wall behind it. A second courtyard then connects the living spaces with the most private bedroom area hidden just behind. This respectful hybrid of work and leisure pays homage to Japanese tradition in the modern era—a temple for tranquility and craft. ∎

PLANS & DRAWINGS

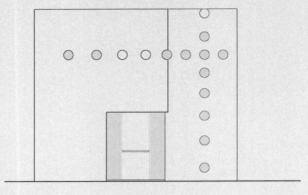

Elevation

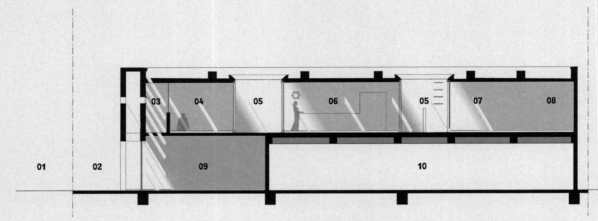

Section

01 Street
02 Shop entrance
03 Hall
04 Work room
05 Sky space
06 Living
07 Bedroom
08 Closet
09 Shop
10 Parking
11 Top of parapet:
    concrete finish
12 External wall: exposed
    concrete finish

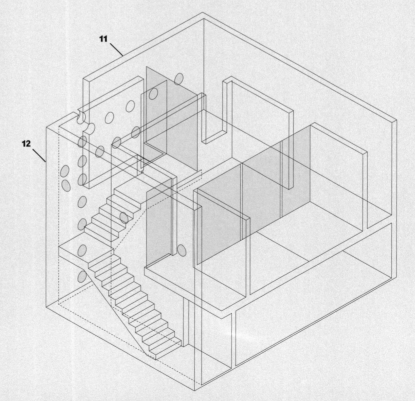

Partial axonometric

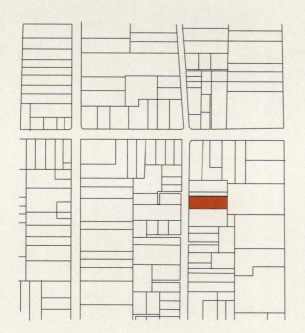
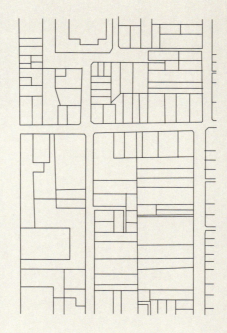

Site plan

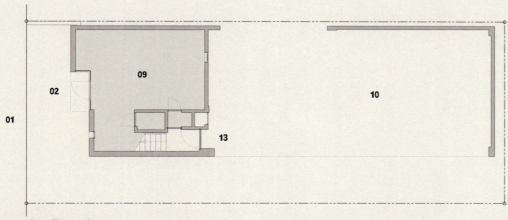

First floor plan

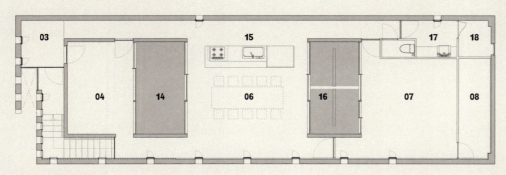

Second floor plan

Spatial Strategy

With workspace at the front and living areas in the back, this house seamlessly integrates both activities. Two courtyards on the upper level bring daylight and fresh air down into the interior, forging a connection between architecture and nature.

●

Number of square meters:
259.78
Number of rooms:
5
Number of floors:
2
Number of residents:
3
Sustainable features:
High standard insulating material
Outdoor areas/rooftop access:
Two courtyards
Context:
City center
Type:
New construction

●
01 Street
02 Shop entrance
03 Hall
04 Work room
06 Living
07 Bedroom
08 Closet
09 Shop
10 Parking
13 House entrance
14 Terrace 1
15 Kitchen
16 Terrace 2
17 Dressing room
18 Bathroom

Takeshi Hosaka Architects
DAYLIGHT HOUSE
Yokohama, Japan, 2011

■ Amidst a densely populated urban area, this transcendent house affords a couple and their two children the opportunity to live under the natural light from the sky. The site, a five-minute walk from the railway station, is surrounded on all sides by individual houses, high-rise residential towers, and offices. Further underlining the less than optimal site conditions, the plot's ground level sits about three meters below the main road, making

the site a literal valley floor among the surrounding buildings. With little to no direct sunlight to be found, the house instead leverages the precious light coming from above.

This luminous home consists of one single-story, multipurpose main room with a high ceiling. To accentuate the family's experience of the expansion of the ceiling and its expressions in each room, the master bedroom, the rooms of the children, and the study are divided by doors and furniture that rise only to

about half the height of the ceiling. Due to the fact that transparent top lights would completely expose the interior when viewed from the surrounding towers, the ceiling instead nests beneath the lighting. This intentional space between the skylights and the ceiling elicits a more subtle relationship to the sky overhead. Rather than observing it directly, the ceiling instead delicately highlights the small changes in the clouds as they shift and glide by during the course of the day.

The ceiling consists of a grid of acrylic vaults. These vaults pull in the movements and atmospheric qualities of the light and sky from the glass skylights, softly projecting them down onto the ceiling. Minimizing the presence of molding, the materiality of the mortar floor and larch plywood walls work together with the detailing of the white acrylic to orchestrate a cohesive and refined interior experience. The effect of these humble materials together produces a level of abstraction where only the light and its nuanced character emerge at the top of the room.

Each room has a rectangular window at sitting height to allow the air to circulate within the living zones of the entire house. The depth of the acrylic ceiling also helps minimize solar heat gain during the summer through forced ventilation. During the winter, this layer of air creates a heat buffer that stabilizes the climate of the interior environment.

WHEN ONE ENTERS THE HOUSE, THE INTERIOR OVERFLOWS WITH SUNLIGHT, MAKING IT DIFFICULT TO BELIEVE THAT THE VIRTUALLY SUNLESS SITE CAN ACHIEVE SUCH A RADIANT QUALITY. True to its name, Daylight House showcases all things related to light and illumination. Not simply focusing solely on sunlight, the

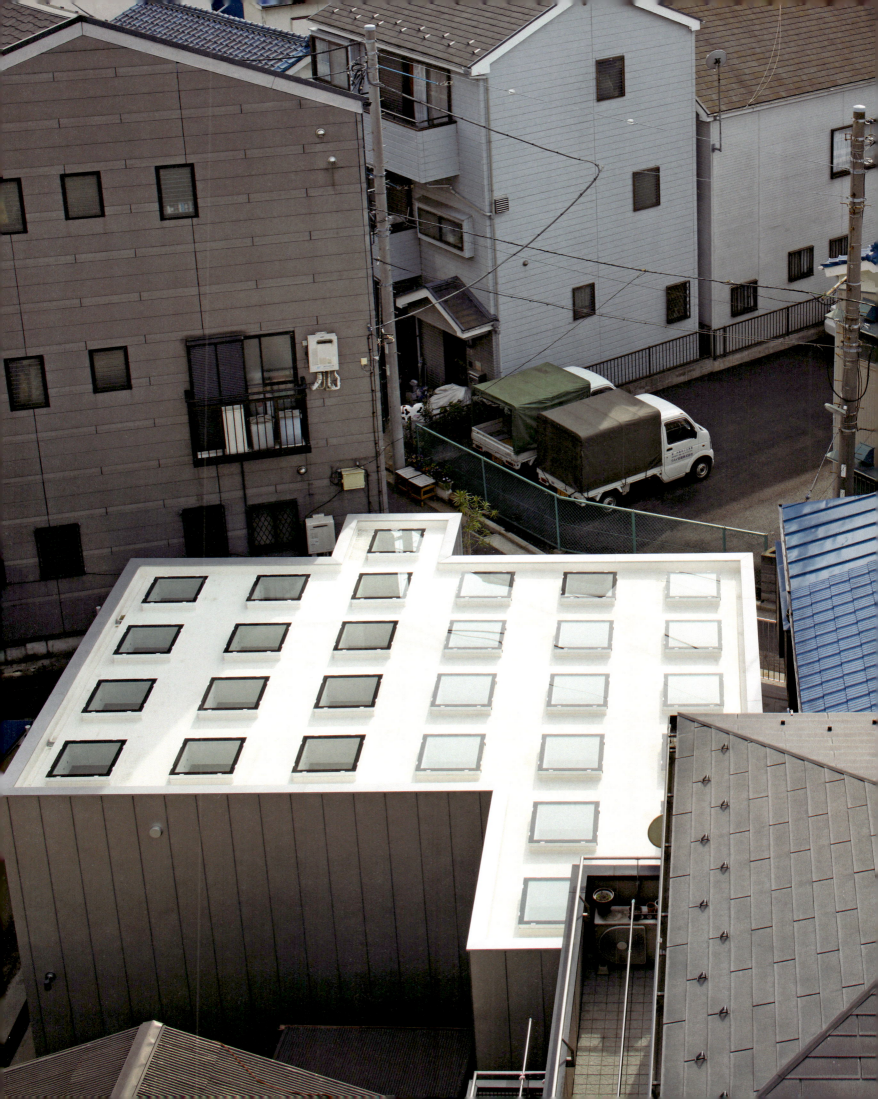

With little to no direct sunlight to be found, the house instead leverages the precious light coming from above.

home responds to the changes in color of the sky at daybreak, how the sun tracks across the sky throughout the day, the passing shadows cast by clouds floating by, the color of the rich hues of the sky at sunset, and how the moonlight softly filters in during the evenings. The admiration of the night sky represents another ritual enjoyed by the family once the lights are turned off at bedtime. This remarkable ceiling of natural light offers the family the luxury of experiencing the transitions of the outdoors around the clock.

Laminated veneer lumber beams (LVL) combine to form a gridded structure that spans the entirety of the roof. Reinforced with plywood, each square of this grid is topped with its own independent skylight. These evenly spaced, modestly sized apertures with Low-E glass ensure an even distribution of light across the interior. For the ceiling below the roof, a curved form made of translucent acrylic is mounted to each square of the gridded beams. These arched pieces generate an elegant vaulted effect and diffuse the light and outdoor scenery from above. Plywood cut into a semicircular shape then attaches to the LVL beams to provide additional support and a clean finish. This dry construction method makes for easy maintenance, as the acrylic can be taken out by removing the screws.

The spacious and airy floor plan is oriented around a central open kitchen, living, and dining area. This shared communal zone capitalizes on the full height of the tall ceiling, granting a lofty ambience to the space. A custom designed glossy table resides in this area. The high gloss of the table inverts the ritual of gazing up at the sky. Instead, by looking deep into the reflective table, the family can carefully take in all the ethereal details of the ceiling above and the nature beyond without needing to crane their heads. Four simple bedrooms and a small study surround the main living area and can be opened up or closed off with folding doors. In spite of appearing lower in height, each of these rooms comes without a closed ceiling. By keeping the top of these rooms open, all spaces are able to enjoy access to the skylights above while still maintaining privacy between one another and the living areas. Small trees rise out of the floor in several locations and infuse the interior with an organic warmth and a direct connection to nature. A flight of stairs and a slender step ladder lead up to an intimate loft. From here, one can lie back and admire the mystical quality of the great beyond. ∎

PLANS & DRAWINGS

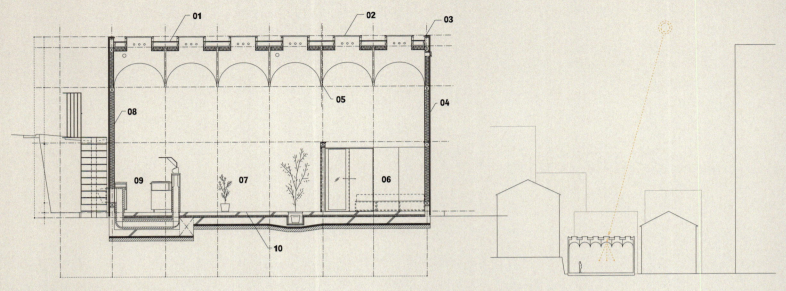

Section 1

Sun diagram

North elevation

South elevation

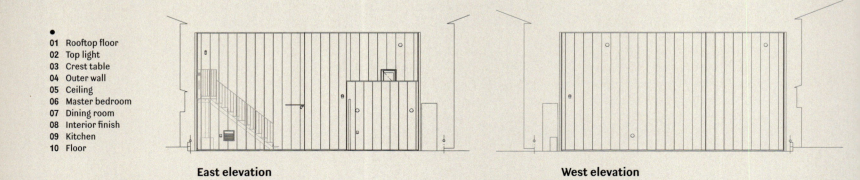

01 Rooftop floor
02 Top light
03 Crest table
04 Outer wall
05 Ceiling
06 Master bedroom
07 Dining room
08 Interior finish
09 Kitchen
10 Floor

East elevation

West elevation

Spatial Strategy

A network of skylights illuminates a dark and narrow site. Conceived as a large open box, the layout features one main room with all private spaces branching off from this central core.

Number of square meters:
73.60
Number of rooms:
8
Number of floors:
1, 2 levels
Number of residents:
2 adults, 2 children
Sustainable features:
Ventilation prevents solar gain during the summer
Outdoor areas/rooftop access:
N/A
Context:
City center
Type:
New construction

Base level

Second level

07 Dining
09 Kitchen
11 Entrance
12 Bedroom
13 Living
14 Bathroom
15 Study
16 Half bath
17 Closet
18 Balcony
19 Loft

Hiroshi Nakamura & NAP
OPTICAL GLASS HOUSE
Hiroshima, Tokyo, 2012

■   A residential oasis for a fortunate family of five rises alongside a busy urban thoroughfare in downtown Hiroshima. Carving out a space for tranquil splendor, the introspective house elegantly blocks out its high-rise commercial neighbors. To achieve utmost privacy and refuge in this chaotic city environment, an optical glass block façade shields the peaceful garden residence from the street. Visible from all rooms, a garden comprised of mature maple, ash, and holly trees frames the different living spaces with nature. This direct access to lush greenery combined with the mesmerizing, soundless scenery of the cars and trams passing by imparts a visual richness to life in the house.

Sunlight from the east refracts through the glass to form ethereal and fluctuating light patterns around the home and garden. A shallow water basin doubles as a skylight, filling the lower entry floor with shifting water patterns from the falling raindrops. Filtered light through the garden trees flickers on the living room floor and a super lightweight metal curtain billows in the wind. Lined with low shrubbery, a secret pathway runs in between the courtyard garden and the glass block façade. This interstitial space allows the family to take a closer look at the happenings of the bustling city just outside when they feel curious. Otherwise the walkway and garden establish an efficient buffer distance between the residence and the outside world. The glass exterior intersects with two monolithic concrete walls on either side of the courtyard. Such a dramatic shift in materials further highlights the transcendent qualities of the delicate façade and the alluring shadows it throws over the courtyard's perimeter.

Despite its location in the city center, the house exudes a rare serenity, affording its residents the unique opportunity to observe the evolving nuances of nature and urban life simultaneously. Applying the traditional Japanese technique of compression and expansion of spaces, the low and wide proportions of the living and dining rooms make the transitions to the outdoor garden areas at the front and back of

the house that much more exhilarating. The layout of this first floor opens up completely on both sides to support a fluid relationship to indoor and outdoor living. A plush floor rug and an L-shaped couch pick up on the teal blue accents of the reflecting pool just a few steps away. Referencing the wooden entrance on the ground level, similar panels wrap around the back wall of the living room and into the dining room. Low

Despite its location in the city center, the house exudes a rare serenity, affording its residents the unique opportunity to observe the evolving nuances of nature and urban life simultaneously.

shelving and a built-in fireplace extend along the outer wall of the space. Shielding the home's main staircase, the placement of a miniature version of the glass block façade brings light into the stairwell. The introduction of the glass exterior within the interior unifies the residence's aesthetic atmosphere. Embracing a tree house quality, the family's bedrooms and large master bathroom on the second floor look out onto a canvas of fluttering leaves.

A FAÇADE OF SOME 6,000 PURE GLASS BLOCKS REPRESENTS THE TOWNHOUSE'S MOST STRIKING FEATURE. Beginning one level up to ensure privacy from foot and street traffic while making space for a garage, the starting point of the glass enclosure corresponds to the first residential level of the house. These pureglass blocks, with their large mass-per-unit area, effectively shut out sound. While the glass cancels out the noise of the city, it still retains its vibrant imagery. The transparent blocks gracefully admit the urban scenery into the clearly articulated garden and adjacent open living and dining rooms. To realize such an intricate façade, a glass casting technique was employed to produce glass of extremely high transparency from borosilicate, the raw material used for optical glass. The casting process proved exceedingly difficult, requiring both slow cooling to remove residual stress from within the glass and strict dimensional accuracy. Still retaining micro-level surface asperities, the resulting network of varied glass bricks generates breathtaking and unexpected optical illusions in the interior and exterior spaces.

The structurally reinforced 8.6 m × 8.6 m glass block wall achieves an entirely transparent effect when seen from the garden. From the street, passerby are greeted with shifting images of tall trees rustling inside the lofty inner courtyard. As both the people outside and the trees inside begin to move, the glass softly blurs the image. Through this array of shimmering looking glasses, the façade oscillates between representation and abstraction—

a crisp photograph and a blurred painting. The prismatic wall results in an intriguing image that never settles. Scattering light over the tastefully appointed modern Japanese interiors of rich woods, light marbles, and matte concrete, the façade appears like a waterfall flowing downward. With access to light, nature, and the city across every space, this remarkable home lives in blissful awareness of the changing seasons and the changing urban fabric. ∎

PLANS & DRAWINGS

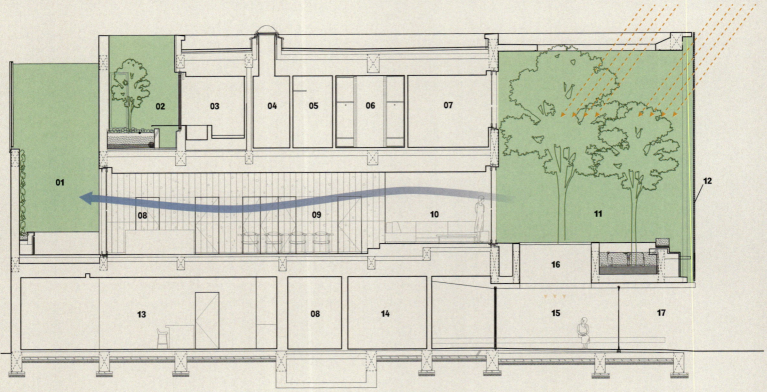

Section

- **01** Ivy garden
- **02** Olive garden
- **03** Bathroom
- **04** Sauna
- **05** Laundry room
- **06** Half bath
- **07** Children's room
- **08** Kitchen
- **09** Dining
- **10** Living
- **11** Iroha Momiji garden
- **12** Optical glass façade
- **13** Hobby room
- **14** Bedroom
- **15** Entrance
- **16** Water basin skylight
- **17** Front porch

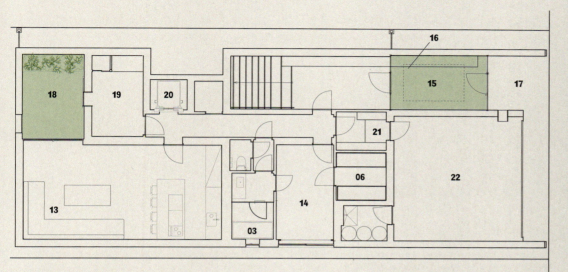

First floor plan

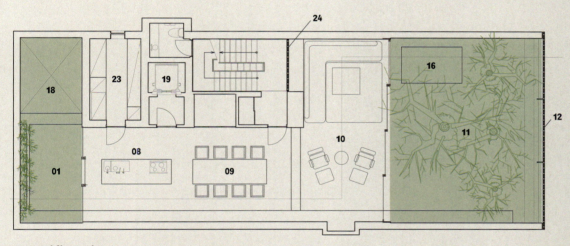

Second floor plan

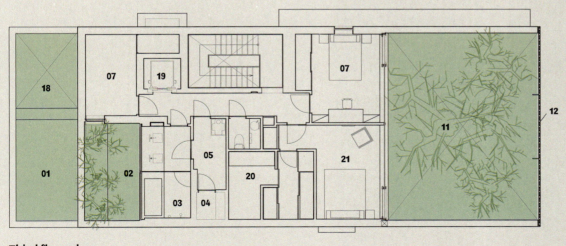

Third floor plan

Hiroshi Nakamura & NAP

OPTICAL GLASS HOUSE

Spatial Strategy

6,000 glass bricks shield a three-story oasis. With a large front garden and a smaller back yard green space, all of the rooms of the main living spaces retain a strong connection to the outdoors and the natural world.

●

Number of square meters:
363.51
Number of rooms:
19
Number of floors:
3
Number of residents:
5
Sustainable features:
Natural ventilation and cooling
Outdoor areas/rooftop access:
Two courtyard gardens
Context:
City center
Type:
New construction

●

01 Ivy garden	15 Entrance
02 Olive garden	16 Water basin
03 Bathroom	skylight
04 Sauna	17 Front porch
05 Laundry room	18 S. Monadelpha
06 Half bath	room
07 Children's room	19 Tatami room
08 Kitchen	20 EV
09 Dining	21 Shoes room
10 Living	22 Garage
11 Iroha Momiji	20 Closet
garden	21 MBR
12 Optical glass	23 Stock
façade	24 Optical glass
13 Hobby room	wall
14 Bedroom	

TERRA E TUMA

MARACANÃ HOUSE

São Paulo, Brazil, 2009

combination of details catches the eye of the curious passerby. The discordant geometry and materials at play stand out in relation to the traditional houses of the neighborhood. **BOTH A PRIVATE RESIDENCE AND A PUBLIC EVENT, THE SURPRISING HOME TAKES ON A COMMANDING ROLE ON THE STREET.** Making the most out of the space available, the property revels in its limitations. This sensitivity to and awareness of the area's context enables the home to internalize the lessons of its surroundings and distill them into a completely unique experience.

MORE THAN AN INTERIOR, THE TWO LEVELS OF THE HOME GRADUALLY FORM A PATH WHERE OUTSIDE AND INSIDE MERGE INTO A PROPER AND CONTINUOUS SHAPE. The house discovers new possibilities for challenging the limitations imposed by the scanty plot. With the lot's complexity exceeding both horizontal and vertical routes, a fresh spatial experience develops. These unusual ways of traversing and occupying space help elucidate the singularities of the district's geography. The house lives in harmony with the neighborhood and its peculiarities. Unraveling expectations, the home and its occupants constantly learn from their diverse surroundings. Whether contemplating the reddish roofs of the neighbors, the church façade which crowns the district, or simply admiring the sunset as it passes over São Paulo's horizon, the house stays finely tuned to the city and the nuances of the rich urban environment.

■ In a city of extraordinary contrasts, a thoughtful indoor/outdoor home finds its place within the urban fabric. The three-story residence appears on the western border of the metropolis. Envisioned as a type of social happening, the house behaves as a new event introduced into the bucolic surroundings. A range of opaque gray materials intersect with clear glass surfaces and a vibrant outdoor mural. Such a youthful and spontaneous

While most houses leave the city outside, this residence retains a deep connection to the world just beyond its walls. Access to the home remains hidden behind a prominently placed ceramic mural painted in black, white, and red compositions. Once inside, the entrance of the house unveils a succession of spaces, that range from the narrow and shady to the generous and bright. This transition leads visitors through a steady stream of new experiences.

Maracanã House

The sequence of the home begins with a double height indoor/outdoor area. Sliding glass partitions and billowy white curtains can demarcate or unify this exterior courtyard and the interior open living room, dining room, and kitchen. From this entry point, a full understanding of the inner workings of the house forms—social and service-oriented programs below followed by more intimate and private areas above. Vines, plants, and flowers make their way into the interior, further promoting a lifestyle enhanced by nature. Similar to the city streets, the lights between spaces transmit in a number of directions through big glass openings. The delicacy and transparency of these windows juxtaposes with the solidity and weight of the cinderblock walls. Modern furniture and wooden accents lessen the rough quality of the concrete and brick finishes. This mixture of untreated materials, large open spaces, and verdant greenery exude an undeniably Brazilian flavor.

Brimming with individuality, the open floor plan produces a circulation route that encourages interpretation. Whether traversing through the spaces or the voids, going out or staying in, the house offers a myriad of ways to occupy it. Residents can relax upon the mass of the ground floor, immerse themselves in the tropical gardens and the lush backyard, or glide up to the second, third, and rooftop levels. At the very top of the building, the roof opens up to the sky. From here, the family can observe the happenings of the city from a bird's eye view.

The house performs as a living infrastructure. Sensitive uses of simple and raw materials allow the architecture to undress the superficial and transitory nature of our reality, elevating only the lasting and meaningful essence of each space. Channeling the humble notions of the primitive hut—the first architectural typology—the pared-down shelter protects from the elements while still reserving a place for nature within the interior.

Maintaining the view of the hillside represents another key factor influencing the design approach.

The hill serves as an important landmark of the Lapa neighborhood. Keeping this vista in mind, the design achieves impressive spans and double-height areas through the investigation and identification of an affordable structural solution. Although a sense of economy influenced the outcome of the residence, the greater desire to create expressive spaces triumphs with the choice of large, high quality windows in feature areas. The resulting home triumphantly addresses the issue of cost without sacrificing the needs of the residents and the quality of the spaces.

Unpretentious and timeless, the home functions as an oasis and place of respite from the bustling city outside. Even so, the residence still nurtures an openness and connection to its context. This blend of private needs and cultural awareness promotes a deeper comprehension of the site and its possibilities. ∎

PLANS & DRAWINGS

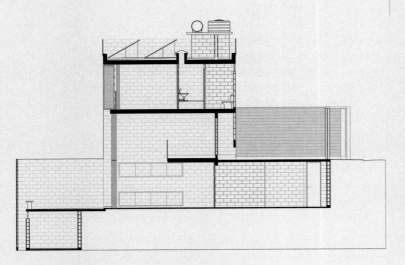

House section 1

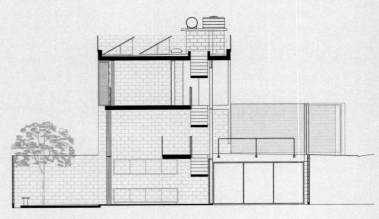

House section 2

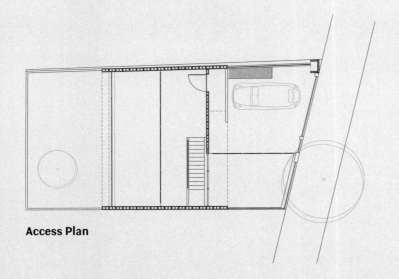

Access Plan

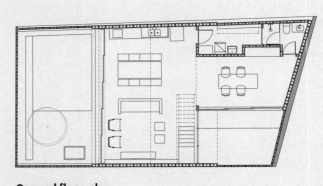

Ground floor plan

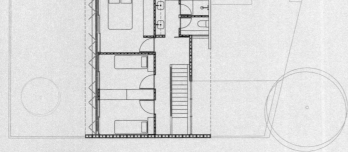

First floor plan

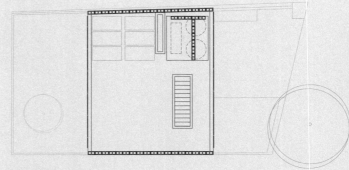

Roof plan

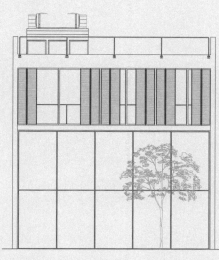

North elevation

West elevation

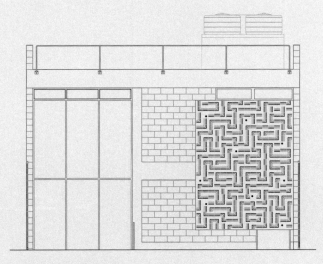

South elevation

MARACANÃ HOUSE

Spatial Strategy

Bright, open, and semi-industrial spaces interweave with lush courtyards. These outdoor spaces appear in both the front and back of the house. The private areas of the house raise up a level to ensure maximum intimacy.

●

Number of square meters:
185
Number of rooms:
4
Number of floors:
3
Number of residents:
4
Sustainable features:
N/A
Outdoor areas/rooftop access:
Courtyard and rooftop terrace
Context:
City
Type:
New construction

■ Designed for a middle-aged newsman, this striking house reflects architectural magazine veteran's keen eye. The site, located on the outskirts of a rapidly expanding area just north of Ho Chi Minh City, gracefully balances individuality with a reverence towards the diverse residential context. Faced with a myriad of contrasting architectural styles, the house still achieves its own unique aesthetic.

Addressing the client's wishes for a special house that doesn't compete with its nextdoor neighbors, the final outcome merges the outdoors with the comforts one expects from the most private interior. A lightweight structure comprised of steel and metal sheets works within the client's modest budget. This structural system stands out against the typical backdrop of the region's brick and concrete constructions. **FINDING ECONOMIC SOLUTIONS THROUGHOUT THE DESIGN PROCESS, THE USE OF RECLAIMED FURNITURE AND FINISH MATERIALS POPULATES THE INTERIOR SPACES.** These second-hand furnishings not only minimize construction costs but also give the house a distinctively appealing look and flexible layout. From vintage chairs to shutters and doors painted in vibrant colors, the new home exudes a beauty and serenity through its repurposed items that typically only comes with old age.

The steel structure not only makes the foundation lighter, but also helps shorten the construction period. By speeding up the building timeline, the home saves costs over all stages of the design process. The framing uses 90 × 90 steel columns and 30 × 30 steel beams connected to metal sheets. Plants and curious vines gradually infiltrate and cover this structural system, making the home appear like a mysterious green box from a distance. Between these cool metal bars, nature begins to define itself as an active character within the house.

Organized into two vertical parts, the layout of the house locates two private bedrooms on the upper floor. Strips of dark wood flooring juxtapose with corrugated metal siding and charming doors painted in a

The residence's spatial strategy considers the house as a group of interrelated nests. Set within a protective perimeter caging, the sizes and shapes of these nests meet the user's personal needs while responding to the constraints of the site. Located among the trees planted all over the site, these areas shield the home from the noisy streets outside. Away from the public eye, this spatial layout allows the private interior activities of daily life to thrive. **THE THOUGHTFUL INCLUSION OF PLENTIFUL GREEN SPACE AFFORDS THE CLIENT THE RARE OPPORTUNITY TO EXPERIENCE AUTHENTIC NATURE INSIDE THEIR OWN HOUSE.**

From the street, the steel frames painted in white take on an intriguing patchwork quality. Mixing scales and transparencies, a larger panel folds open to welcome the city inside. Salvaged floor tiles spill out to the edge of the entry and across the entire ground level. Moving from the courtyards to the walls of the kitchen island, and past the living room, this tapestry of brilliant colors and graphic patterns culminates at the end of the backyard patio. The mesmerizing sea of painted tiles adds to the tropical and timeless feeling of the house. Locally made, these tiles build on the recycled materials concept present throughout the design. The benefits of these tiles are that they not only reduce material and finishing costs but also elegantly transform the useless into the meaningful.

In addition to the organization of spaces and the flexible uses of structure, the housing concept demonstrates a low cost solution with priceless visual and

range of cheerful tones. Old-fashioned shutters, each boasting a bright shade of orange, yellow, or green, create a youthful flare. A delicate white stairs leads up to a covered rooftop terrace that looks out into the garden below. Lemon yellow and cherry red curtains slide across the terrace, lending it a more intimate atmosphere as needed. The garden level on the ground floor consists of an open kitchen and living room that sweeps out onto a verdant backyard patio. **THE TRANSITION FROM INDOOR TO OUTDOOR LIVING OCCURS WITHOUT THE USE OF A SINGLE BARRIER, DOOR, OR WINDOW.** This direct and constant connection between interior and exterior spaces effectively erases any distinction between the two. Limiting the footprint of the interior living area, the floor plan instead devotes its attention to the spacious garden area just steps away. This wild and lofty outdoor area amplifies the spatial experience and threshold between inside and out. The introduction of this untamed nature serves as the main focal point of the house. Influencing and linking to each interior space, the trees act as the building's secondary walls.

spatial results. The home with its dazzling aesthetic economy not only attracts the attention of the neighbors but also instigates a rethinking of the traditional Vietnamese urban dwelling. Minimizing the budget without sacrificing on beauty, this project interweaves nature, privacy, economy, innovation, and recycling into euphoric architectural expression. ∎

PLANS & DRAWINGS

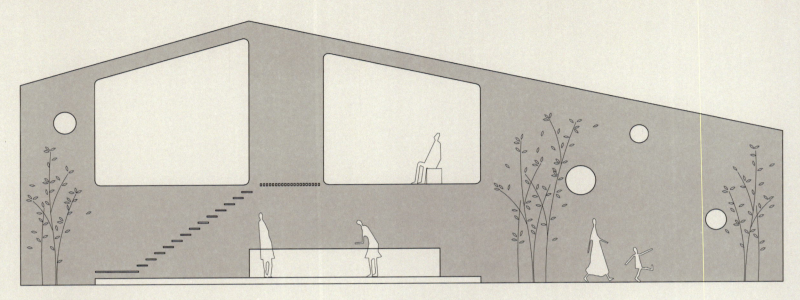

Longitudinal section

Sectional diagram

THE NEST

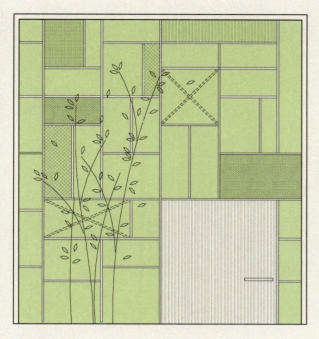

Elevation

Spatial Strategy

Generous front and back courtyards arrange a fluid indoor/outdoor lifestyle. These garden spaces flow into an open kitchen, living and dining area with the more private bedrooms of the house situated one level up.

●

Number of square meters:
200
Number of rooms:
2
Number of floors:
2
Number of residents:
1
Sustainable features:
Use of lightweight materials reduces construction cost and time, salvaged furniture and building materials, natural ventilation, trees and plants act as sun screens
Outdoor areas/rooftop access:
Large outdoor garden patio
Context:
Residential area
Type:
New construction

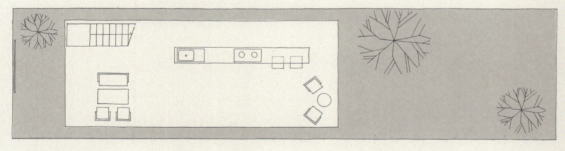

Ground floor plan

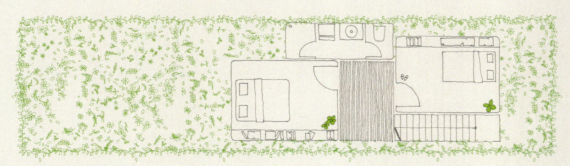

Second floor plan

CASE-REAL
WHITE DORMITORY FOR IL VENTO

Teshima, Japan, 2011

history and traditions are carefully preserved. One instance of this preservation appears in the untouched existing exterior wall facing the street lane. Despite replacing the deteriorated but traditional Japanese style roof, the basic form and color of the original inspires the design approach of the replacement structure. This updated roof uses different materials from its predecessor but still manages to read as relatively unchanged from the outside. The contrast between old and new occurs in the choice of all the furnishings and functional features of the dormitory and courtyard. These elegant components are reconstructed with various white materials such as woods, stones, and plasters that offer a range of expressions and tones. **IN JAPAN, WHITE SYMBOLIZES NOT ONLY A NEW BEGINNING BUT ALSO STANDS AS THE SACRED COLOR REPRESENTING PURITY, INNOCENCE, AND PEACE.**

The interior space comprises three private rooms with individual cooking areas as well as a shared lounge with a small kitchen. An elegant courtyard sits in the middle of these residential structures. Hovering just above the ground, a sprawling white terrace links to the lounge and imbues the open space with a sense of relief and freedom. This raised terrace winds across this outdoor area in a U-shape. A single tree rises from the circular cutout placed at the back of the deck. From the outside, a series of square stepping stones leads visitors down a narrow alley into the core outdoor area of the house. This slender path faces the original perimeter wall composed of an assortment of different scales of stone. Shifting from the coarse and rugged to areas of more fine and intricate stonework, the wall serves

■ A renovation project converts an old vacant house into a dormitory in a village on Teshima, an island lying in the Seto Inland Sea. Created as an artwork by Tobias Rehberger for Setouchi International Art Festival 2010, this serene dormitory houses three members of IL VENTO's restaurant staff. The main objective for this residence focuses on how to make the most of the vacant houses left behind by the increasing depopulation of the region.

The humble architectural response tackles the challenge of inserting new design elements without disturbing the surrounding landscape of the village. Remaining an unchanged microcosm, the village's

White Dormitory for Il Vento

as a reminder of the deep historical roots embedded throughout the site. Referencing the curvature of the terrace, the stepping stones gradually evolve into a paved entry path that establishes a smooth transition from the street to the meditative spaces within.

Contrasts between the inside and outside of the building, or in other words the past and the present, expand the possibilities of the existing architecture in the neighborhood and on the island itself. Sliding glass doors extend the lounge area out onto the terrace. This seamless transition between indoor and outdoor spaces gains further support due to the continuation of the simple interior benches made from a light wood. Softened by cushions, these benches run across the back wall of the lounge and halfway onto the terrace outside. Gauzy white curtains act as ephemeral boundaries. These curtains close off or open up the inside and outside areas depending on the mood and need of the residents as they react to the various spaces and each other. The ethereal white lounge space features a strip of skylights directly above the indoor section of the bench. Spanning from wall to wall, these skylights behave as a sleek design tool for evenly filling the space with light. **THE ORIGINAL EXPOSED WOODEN CEILING BEAMS INTRODUCE A RUSTIC FEELING INSIDE THE ALL-WHITE LOUNGE.** These organic structural members retain the rough quality and charming imperfections of the tree trunks they originated from. Haphazardly overlapped, the dark and textured beams contrast with their perfectly polished and bright surroundings.

Sustainable repurposing takes place throughout the house. In addition to the salvaging of the extant wood frames and the exterior wall, the striking original wooden beams in the lounge are also left untouched. These elements consolidate into a pristine and modern rectangular box topped by a traditionally inspired, pagoda-shaped roofline.

Each of the three apartment units consists of a light-filled space with modest furnishings, built-in shelving, and desk space. All of the cozy, private rooms come with their own personal kitchens and bathrooms. In spite of the limited space, these studio units provide a sense of privacy and airiness without limiting the amount of light exposure into the room. The spaces efficiently utilize every nook and cranny, producing a feeling of intimacy in sync with their tailored qualities. Both the indoor and outdoor areas defy any feelings of claustrophobia within these narrow corridors and compact layouts. Such thoughtful spatial planning underlines a sense of directionality as the interior areas pick up where the exterior leaves off. A hybrid outcome between old and new styles and traditions, this residential project masters the art of living well and within one's means. ∎

PLANS & DRAWINGS

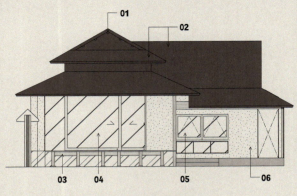

South elevation (courtyard)

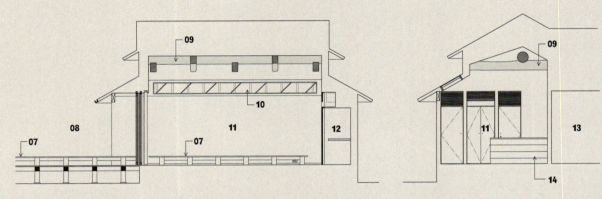

Elevation 2, section

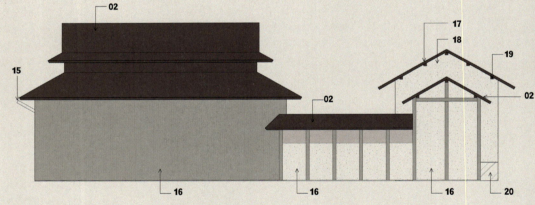

West elevation

-
01 Existing purlin
02 Galvalume roof (new)/ same form and color with the original roof
03 Cedar deck (new)
04 White frame sliding door (new)
05 White frame window (new)
06 Plastered white wall (new)
07 White dyed bench (new)
08 Terrace
09 Existing beam and purlin
10 Top light (new)
11 Lounge
12 Storage room
13 Room 1
14 Kitchen counter with white dyed red cedar (new)
15 Existing light
16 Existing wall
17 Existing Japanese tiled roof
18 White plastered wall (new)
19 Existing purlin
20 White galvalume plate (new)

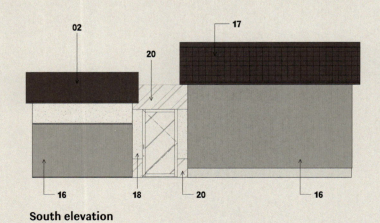

South elevation

WHITE DORMITORY FOR IL VENTO

Spatial Strategy

Three modestly sized dormitory units open up to a large, U-shaped outdoor deck. This central deck connects directly to a shared indoor lounge space with a communal kitchen.

●

Number of square meters:
94.2
Number of rooms:
3
Number of floors:
1
Number of residents:
3
Sustainable features:
Repurposed materials and structural elements
Outdoor areas/rooftop access:
Courtyard
Context:
Village
Type:
Addition and renovation

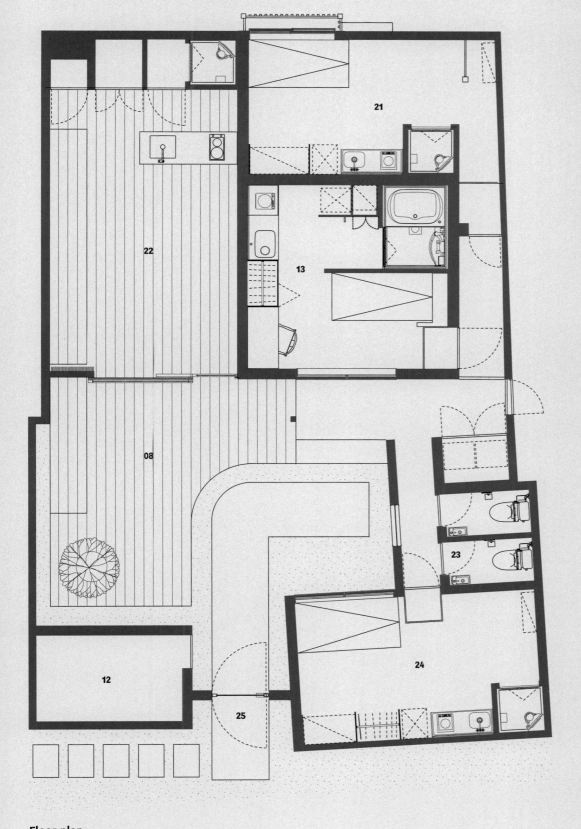

Floor plan

●

08	Terrace
12	Storage room
13	Room 1
21	Room 2
22	Lounge space
23	Half bath
24	Room 3
25	Entrance

Edwards Moore
DOLLS HOUSE
Melbourne, Australia, 2013

A humble renovation of a worker's cottage also doubles as the smallest house on the street.

The updated residence retains the existing street frontage and primary living areas while fragmenting the building's addition behind. The clever expansion introduces courtyards that provide separate yet connected functions for living. Collections of raw and untreated finishes create a rugged ambiance that compliments the owner's desire for an uncomplicated living arrangement.

Echoes of the home's history are reflected in the discreetly choreographed gold panels located throughout the space. The abundance of natural light refracting off the white interiors channels a sense of the

ethereal—an otherworldly environment hidden amidst the urban grain. A refreshing blend of old and new, the home resembles an elegantly repurposed ruin.

The modest residence inverts the traditional relationship of house and garden. Instead, the layout extends the house the entire length of the site by inserting two courtyard garden spaces back into the house. The placement of these courtyards grants every room an external aspect or access to one of the four key spaces—kitchen, living area, bedroom, and bathroom. Enjoying southern exposure, these gardens soak up the sun throughout the afternoon. Filled with coarse gravel, the angular courtyards are backed by the brick walls of the adjoining building. These brick walls not only produce an industrial character but integrate metal trellises to hang potted plants and encourage their vines to grow.

WORKING WITH A SITE UNDER FOUR METERS WIDE, THE LONG AND SLENDER LAYOUT DOES A REMARKABLE JOB OF ENHANCING THE FEELING OF SPACIOUSNESS THROUGHOUT. The front of the house, guarded by the original unpretentious white picket fence, reveals few clues about the demure paradise awaiting just inside. Sharing walls on either side with the two next-door neighbors, the edges of the rough brickwork are softened by wild vines that creep up and over the conjoined homes.

After passing through the old-fashioned, black front door, the house opens up to reveal a series of hybrid indoor and outdoor spaces. Unfinished walls, exposed bricks, and spray-painted construction markings purposefully remain as tactile memories of the building's history and story. Preserving the original house, the layout begins with a narrow living space. This charming living room features floor to ceiling bookshelves and a fireplace with original molding. A rickety stairs descends from a loft area above, beckoning the curious. Minimalist shelving frames the old front window, adding flexible storage for books, records, and a rotating assortment of the client's belongings. The playful classic coat hanger the Eames lines the back of the entry door.

The all-white space takes on a mesmerizing glow as daylight spills in from the courtyard. A corridor leads to a small bathroom nestled in the back corner of the space. Glass doors framed with a rich wooden trim allow the different spaces to connect both visually and physically to the outdoors. Similar to the home's open floor plan, where one room meanders into the next, the courtyards act as extensions of the interior—peaceful havens of tranquility. Separated by the second courtyard, the bedroom, kitchen, and dining space perfectly align. The interconnected relationship of these distinct spaces is enhanced when both sets of courtyard doors are opened.

IN ADDITION TO THE HOME'S COMMITMENT TO THE PAST, IT ALSO UTILIZES A NUMBER OF FORWARD-THINKING SUSTAINABLE ELEMENTS. These sustainable features include rainwater harvesting and a gas water heater. Part ruin and part temple, the juxtaposition of past and present choreographs an uplifting way of relating to both indoor and outdoor spaces. Here, old world arched doorways with original moldings intersect with contemporary detailing and weathered finishes reminiscent of an industrial warehouse. This visceral palimpsest of styles and periods orchestrates a timeless and nuanced method for interacting with history and one another. ■

A corridor then leads to the U-shaped kitchen which opens up to a courtyard on each side. Designed to match the materials of the extant part of the house, the timber-floored kitchen and dining room physically and visually engage with both the outdoors, and the front and back living spaces. The dining area faces an excavated brick wall with a large fragment of sheetrock left behind. Adorned by a simple wooden square frame holding a skull, this unpolished leftover finish material behaves as an impromptu feature wall. A light bulb hangs down over the table encased in a geometric wooden enclosure. This sculptural lighting fixture plays off the idiosyncratic mixture of styles and finishes appearing in every corner of the house.

From there, the layout then leads to the back courtyard and airy bedroom tucked behind. This intimate room makes the most of its pitched roof ceiling. A single lamp hangs down from the highest point of the ceiling, illuminating the cozy white bed below.

PLANS & DRAWINGS

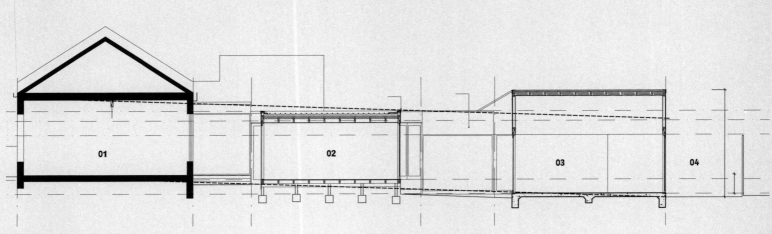

Section 1

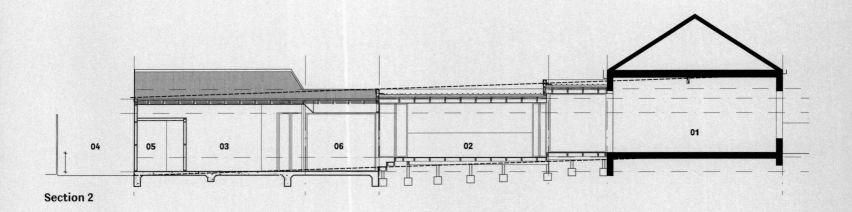

Section 2

- 01 Living
- 02 Kitchen/dining
- 03 Bedroom
- 04 Laneway
- 05 Half bath
- 06 Corridor

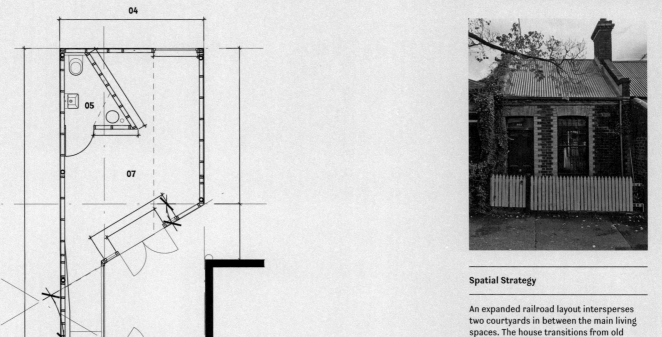

Spatial Strategy

An expanded railroad layout intersperses two courtyards in between the main living spaces. The house transitions from old to new, with the most recent and private additions appearing in the deeper parts of the site.

Number of square meters:
66
Number of rooms:
4
Number of floors:
1
Number of residents:
2
Sustainable features:
Rainwater harvesting,
gas water heater
Outdoor areas/rooftop access:
Two courtyard gardens
Context:
City center
Type:
Addition and renovation

04 Laneway
05 Half bath
06 Corridor
07 Bed 2
08 Kitchen (timber flooring)
09 External courtyard
10 Existing house

Floor plan

Level Architects
SKATE PARK HOUSE
Tokyo, Japan, 2011

■   A seemingly inconspicuous Tokyo townhouse in a quiet residential neighborhood of Shibuya reveals a playful and bespoke character just inside. The one-bedroom residence caters to the special interests of a young married couple. Rather than incorporating additional bedrooms into the floor plan, the couple instead opts for more space for their individual passions: skateboarding and piano.

The mixed-use studio space with an indoor skateboard bowl embedded into the floor features multiple angles for diverse interaction and sport. Balancing the roles of studio, exercise space, and entry point, the layout showcases a number of quirky and endearing touches. Built-in shelving outfitted with vintage glass doors stores the shoes of the couple and guests as well as a stray bowling pin, taxidermy pheasant, and an assortment of vinyl records. Sitting close by, an abstract white tree stump supports any foot struggling to put on its shoe. A bright orange carpet with exotic patterns enlivens the taupe-colored space. Ideal for performance, overhead lighting set on a series of rails affords a theatrical quality to the shared rooms.

Raised half a meter off the ground on a brick platform, an intimate piano room resides at the back of the studio. The elevated level of the rehearsal space helps with soundproofing as well as creating a grand, stage-like atmosphere for performance. Acting as a further buffer for sound, floor-to-ceiling wood shelving lines the back wall of the rehearsal room. When the doors open up onto the studio, the expanded space with the skateboard bowl transforms into guest seating, exalting a mere practice room into a

public concert hall. **THE UNCONVENTIONAL JUXTAPOSITION BETWEEN SKATEBOARDING AND PIANO PLAYING SETS UP AN ENGAGING VISUAL AND AUDITORY DIALOGUE BETWEEN THE DISTINCT PROGRAMS.**

Without the need for a car park on the site, the design instead introduces a private entrance courtyard. The sliding glass panels of the first floor open up onto this enclosed area and allow for the

workshop and studio to expand outwards towards the city. This courtyard space pulls the house away from the street to add a second layer of privacy to the living quarters. Shielded from prying eyes, the exterior of the house reads as two solid concrete cubes. A single black door punctuating the ground floor volume serves as the sole adornment visible to passersby. The upper levels, placed several meters back, open up to the courtyard below and city beyond by integrating large windows and skylights into the façade.

After entering through the main courtyard and passing by the studio and skateboard bowl, the first three steps of the home's primary staircase lead up to the piano room. From the practice platform, the stairs seamlessly pick up again and flow into the living spaces above. The living, kitchen, and dining areas utilize a similar concept of half-level changes to both separate and combined programs across the second floor of the house. Shifts in ceiling height and choices of material emphasize the boundaries between the kitchen and dining area and the luminous, high-ceilinged living room a few steps down. Heavy concrete countertops and rich wooden surfaces lend the well-appointed kitchen a touch of drama. A large rectangular cutout breaks open the concrete wall above the sink to bring the light and activity of the stairwell into the kitchen. Similar material approaches and apertures extend into the bathroom and laundry room. A no-nonsense dining space holds a light blue table that references the feature color regularly appearing around the house.

The continuous presence of light throughout the entire residence stands as another method for connecting the various spaces and levels. Accomplished with the installation of a generously sized and centrally placed skylight, daylight not only floods the main staircase but also spills onto each of the three floors. Artworks casually lean against the walls to inspire an informal and cozy ambiance. Exposed wooden beams, rugged concrete, salvaged timber, rough brick work, and unifying accents of dusty teal blue grant the home a rustic yet contemporary quality.

The top level consists of a private master suite and bathroom. In order to create a break from the lower levels, the scale of the materials used here increases accordingly. The many overlapping layers of floorboards produce an ambiguous break between the rooms. This gradual transition and visual interconnectedness from room to room facilitates an open and spacious feeling across the suite. Running across the floors and over the walls, both old and new floorboards spread out past their anticipated boundaries to emphasize a cohesive aesthetic language. **THE BALCONY, IMAGINED AS AN INTERIOR GARDEN, OFFERS AN INTERSTITIAL SPACE FOR PAUSE AND RESPITE BETWEEN BATHROOM AND BEDROOM.** With broad views into the tranquil, deep blue tiled bathroom on one side and the minimalist bedroom on the other, a secondary stair—comprised of sleek floating risers—links to a roof deck one flight up. The exhilarating rooftop terrace clad in wood provides a dazzling panorama of the metropolis—a satisfying culmination for this private oasis. ∎

The continuous presence of light through-out the entire residence stands as another method for connecting the various spaces and levels.

PLANS & DRAWINGS

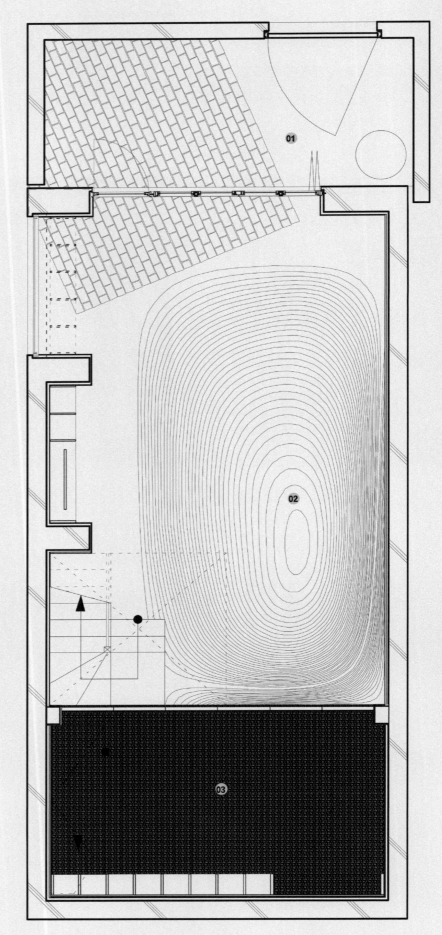

01 Entrance/inner courtyard
02 Studio skateboard bowl
03 Piano room

Ground floor plan

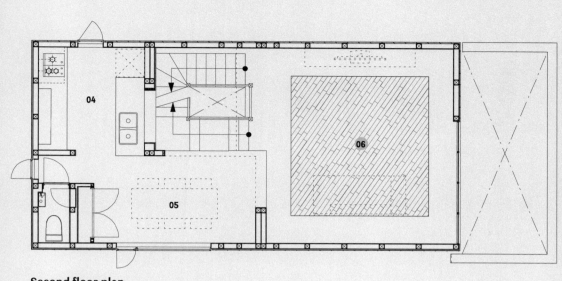

Second floor plan

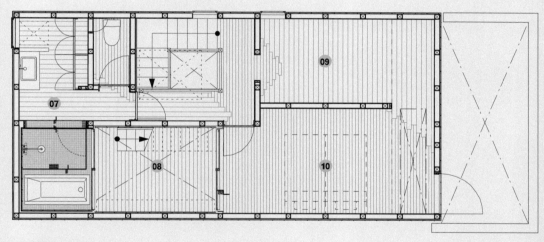

Third floor plan

Spatial Strategy

Opting for personal interests over extra bedrooms, this house makes way for a couple's individual interests in skateboarding and piano playing. These hobby spaces take up the ground floor while the more private areas of the house move upwards through a series of stepped levels that culminate in a rooftop terrace.

Number of square meters:
149.16
Number of rooms:
7
Number of floors:
3
Number of residents:
2
Sustainable features:
N/A
Outdoor areas/rooftop access:
Rooftop balcony and entry courtyard
Context:
City center
Type:
New construction

04 Kitchen
05 Dining
06 Living room
07 Master bathroom
08 Terrace
09 Walk-in closet
10 Master bedroom

Moonhoon
LOLLIPOP HOUSE
Giheung-Gu, Republic of Korea, 2010

True to its name, a pink and white swirl of a home accommodates a young couple and their daughter. The whimsical and hard to miss design stands as the result of the family's request for a unique house, something they had never seen before. Executed on a tight budget, the design

develops a split-level layout of staggered floors organized around a bright central atrium and wrapped up in a striking outer skin. With only one close neighbor at the back of the building, the high visibility of the three main faces of the house enables the structure to take on a commanding role in the community.

WHAT READS AS A PSYCHEDELIC BOX FROM THE OUTSIDE TURNS INTO A HIGHLY EFFICIENT AND ORGANIZED INTERIOR SPACE, CARVING A NOOK FOR EVERY NEED. In spite of its hyper-graphic and showy exterior, the home's interior agenda proves the pinnacle of minimalist serenity. Illuminated by a generous main skylight, the airy whitewashed spaces harmoniously

balance with light wood floors, stairs, and interior accents. The main light well also creates a current that circulates air across the house. The multiple interconnected floors provoke a sense of interdependence and relationship between spaces through intersecting vertical views of both upper and lower levels. A slender wood handrail, reinforced by a delicate, white gridded guardrail below, winds along the stairs and serves a unifying detail over the multiple floors. From the landing at the very top of the stairwell, the angled skylight presents a snapshot of the city's urban density and nearby high rises.

The home's interior consists of a series of half floors revolving around the staircase. Instead of embracing a conventional three-story layout, the multiple in-between levels allow the house to behave as a seven-story building. This dynamic arrangement begins with a half basement that doubles as a study. The basement then vertically transitions into a living room, kitchen, and dining space. Continuing the ascent, the upper levels hold the master bedroom and child's bedroom, with an attic play space and Audio/Visual room at the top. Spiraling around its prominently accentuated stairwell, the house takes on an energetic quality.

THE SPIRALING ENERGY FOUND IN THE INTERIOR DIRECTLY INSPIRES THE SWIRLING EXTERIOR, EVOKING THE PLAYFUL AND VIBRANT AESTHETICS OF AN OVERSIZED LOLLIPOP. Magenta swirls run across the front and back façades of the house. The iconic presence of these two feature walls corresponds to the clients' wish for a house that radiates an intense and sensational identity throughout the neighborhood. Comparatively tame, the same magenta hue extends onto the cladding of the side elevations to establish a consistent exterior language.

Dark grey window and doorframes emphasize all entry points for people, light, and air. These smokey accents stand out from the hypnotic language of the striated façade. Outfitted with a protective overhang, a demure grey staircase invites guests up onto the elevated entry platform at the front of the house. This same grey color palette appears on the exterior guardrails running along the raised main entrance and the upper balcony. Small square windows punctuate the rear façade, maintaining privacy while still letting in the sunlight.

The hot pink and white façades step back on one side to make room for a compact balcony on the second story. From this elevated vantage point, the family can admire the changing atmosphere of the city. The act of lifting all direct contact between the city and

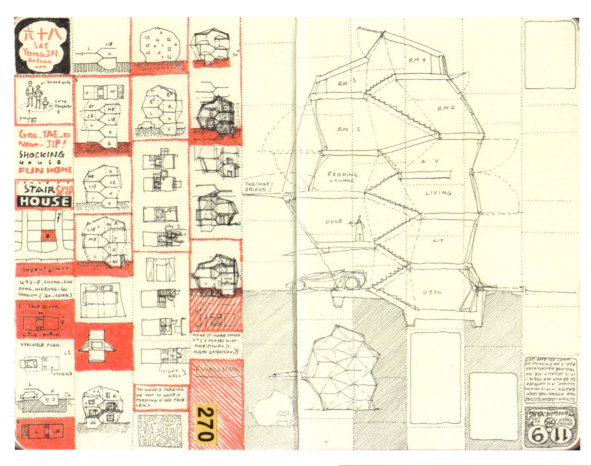

the house up a few meters introduces a level of discretion and privacy for the interior realm. A back entrance connects to the outdoor parking space. This covered entry point extrudes out from the façade to make a transitional room between the outdoors and the main core of the home. Camouflaging the space behind, the same striped patterning folds down and over the extension. By aligning the striped patterns, the extension fluctuates between blending into the main façade and popping out again when viewed from certain angles. This shifting appearance results in a fluctuating reading of the building as it expands and contracts while walking around it.

CONTRASTING WITH OTHER KOREAN BUILDINGS IN MORE WAYS THAN ONE, THE LOLLIPOP HOUSE ALSO DIVERGES FROM THE NATIONAL STANDARD OF CONCRETE CONSTRUCTION. The home instead engages wood platform framing. Both lumber and building method are acquired from Canada, allowing the entire building to be completed in a matter of weeks. This considerable shortening of the construction timeline drastically reduces costs while the wood and metal members make for easy recycling and demolition. The wood studs provide space for proper insulation, something regularly omitted in traditional concrete structures. When applied to a Korean home, the insulation combined with radiant floor heating produces considerable energy savings and a stable indoor climate. Bold yet pragmatic, this memorable residence bridges the gap between efficiency, artistic expression, and economy. ∎

PLANS & DRAWINGS

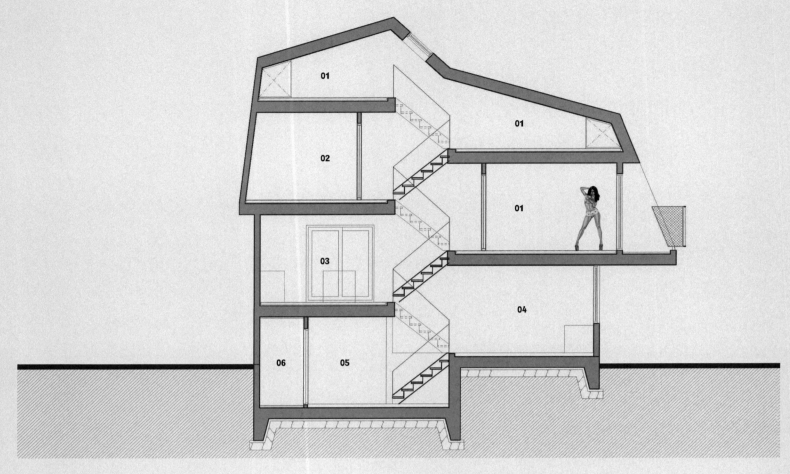

Section

01 Room
02 Child's room
03 Kitchen
04 Living room
05 Study
06 Bathroom
07 Balcony
08 Entrance
09 Storage

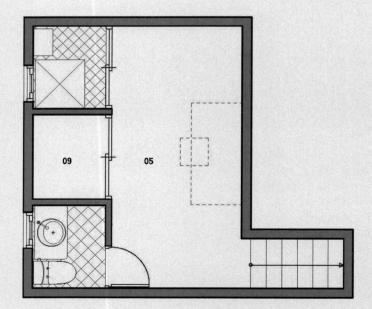

Basement plan

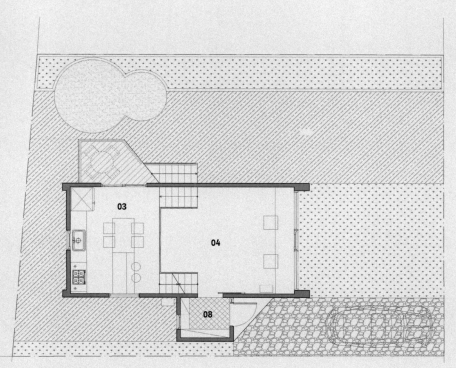

First floor plan

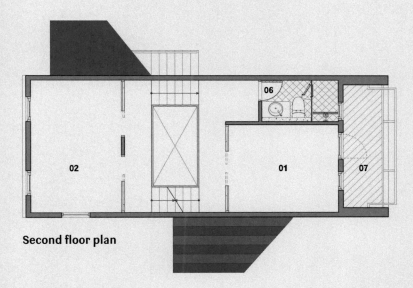

Second floor plan

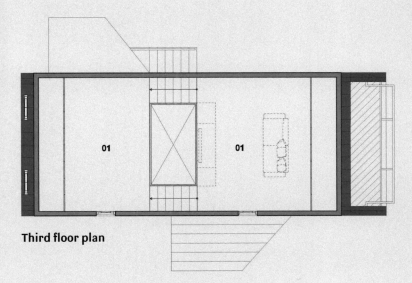

Third floor plan

Moonhoon
LOLLIPOP HOUSE

Spatial Strategy

A swirling floor plan reveals a series of split-level floors connected by a spiraling staircase. By introducing multiple levels, the interior develops a dynamic energy and relationship from space to space.

●

Number of square meters:
102.98
Number of rooms:
6
Number of floors:
4
Number of residents:
2 adults, 1 child
Sustainable features:
Efficient insulation,
made from easily recyclable materials
Outdoor areas/rooftop access:
Balcony
Context:
City center
Type:
New construction

●
01 Room
02 Child's room
03 Kitchen
04 Living room
06 Bathroom
07 Balcony
08 Entrance

Andrew Maynard Architects
VADER HOUSE
Fitzroy, Australia, 2008

■ Emerging from behind its high boundary wall, a dramatic extension to a Victorian terrace home interrupts the symmetrical roofline typical of this Melbourne suburb. The introduction of an articulated, peaked roof over the updated residence breathes new life into the home and the dense innercity of Fitzroy. By distorting a classic pitched-roof silhouette, the memorable roof shape produces a lofty atmosphere below as it gains in height.

The extension consists of a framed steel skeleton. This supporting structure envelopes the unusually high masonry boundary wall, built prior to height restrictions, and reclaims it as a rugged feature wall for the interior. Responding to the imposed site setback, the proud roof doubles as a masterful subverted answer to the rigid regulations. These setbacks allow for double height ceilings, a mezzanine level, and a spacious interior.

The eastern and western façades of the extension are encased in a shield of dark louvers. These thin shades peel back to reveal a dynamic internal environment framed by exposed steel beams and balanced by soft color schemes. Playful splashes of deep red enliven the interior. These broad internal surfaces are occasionally punctured by windows that spread a dancing, cinematic light over the internal workings of the house. Wrapped in the heavy roof form, the refined material and colorful palette of the extension distinguish it from the dark masonry clad terrace from which it emerges. These two opposing forms are united by a transparent glass corridor along the northern boundary wall that frames an outdoor courtyard.

STRATEGIC PLANNING LOCATES THE COURTYARD AT THE CENTER OF THE SITE BETWEEN THE NEW AND OLD STRUCTURES. This prime location grants both the extant part of the home and the extension direct

contact with the outdoors. An elegant plunge pool sits at the core of this outdoor space, affording a tranquil ambiance and natural cooling during warmer days. Fluidly linking the interior with the exterior, flexible sliding glass partitions transform this courtyard space into a central communal zone with

a connecting walkway that can be opened up completely in summer months or return to being part of the interior during cooler seasons. A social magnet, this multipurpose courtyard attracts the activities from the surrounding living spaces, encouraging them to spill out into the open air. The compelling nature of this courtyard ensures that the entirety of the site remains fully utilized. Such a seamless relationship between internal and external environments makes a distinction between the two nearly impossible. Transparent folding doors facilitate constant physical and visual interaction between these two environments, resulting in an extension that is both inside and out. The courtyard's location also

later reveals itself to be one of complexity and ambiguity. Presenting an unassuming face to the street, the back entry to the new addition comprises a staggered brick wall punctuated by an industrial-size roll-up door and a slender glowing window of the same height. From across the street, the prominent roofline gradually comes into view and presents selective glances into the addition's second floor mezzanine and study. The stepped levels of the brick façade seamlessly merge with the steel structure and rhythmic pattern of the louvers above, generating an eye-catching visual composition filled with material contrast. Many elements of the design also serve multiple functions. The bold, crimson red staircase becomes part of the kitchen joinery, the louvers act as light control as much as privacy screening, and the boundary external wall doubles as the internal kitchen wall.

The extension is created out of components that appear to have fallen at the eastern end of the site in a Tetris-like manner. One of these Tetris pieces lodges itself deep within the walls of the original building during its fall. This floating block provides the master bedroom with a tiled en-suite that reflects the rich red and pink accents scattered throughout the interior.

The covered indoor/outdoor walkway between the two sections of the house fuses the original section to the extension. Always offering broad glimpses of what one will experience while moving across the residence, the anatomy of the home also extends far below the site. The timber deck in the courtyard becomes a retractable deck when pulled aside, exposing a hidden spa, right at the heart of the house. Similarly, the timber floorboards in the extension form a trapdoor that uncovers a cellar spanning a considerable distance underground. When these doors are opened, they significantly alter the nature and perception of the home. This adjustable way of using and experiencing the extension gives it a chameleon-like quality, inspiring its occupants to take on an active role of discovery as they engage with and shape their surroundings. ∎

The open and seemingly simple nature of the house later reveals itself to be one of complexity and ambiguity.

provides abundant natural light and ventilation into both the original house and extension. This consistent supply of light and air decreases the family's reliance on mechanical heating and cooling systems. The open and seemingly simple nature of the house

PLANS & DRAWINGS

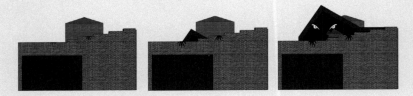

Diagram of the stealth addition

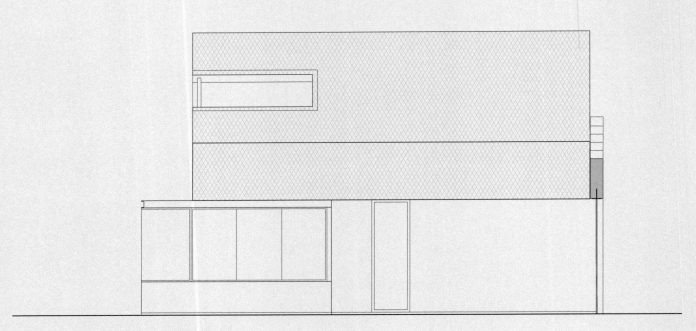

South elevation

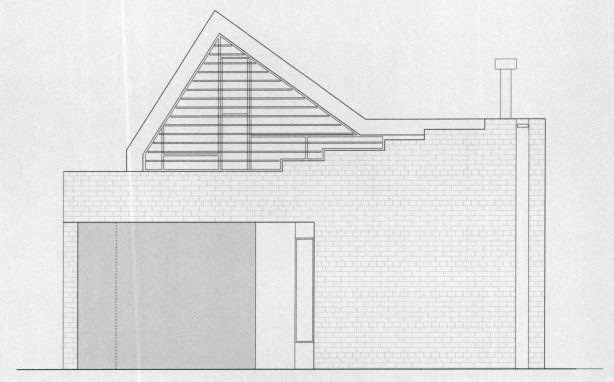

East elevation

Spatial Strategy

A courtyard with a small central pool connects the original house with the backyard extension. The two-story addition grows over the top of the original brick perimeter wall. Sharply angled, the roof makes space below for a lofty living area.

●

Number of square meters:
155
Number of rooms:
8
Number of floors:
2
Number of residents:
1
Sustainable features:
Double glazed windows,
automated louvers for shading
and climate control,
high performance insulation,
rooftop solar panels
Outdoor areas/rooftop access:
Courtyard
Context:
Inner suburb
Type:
Addition and renovation

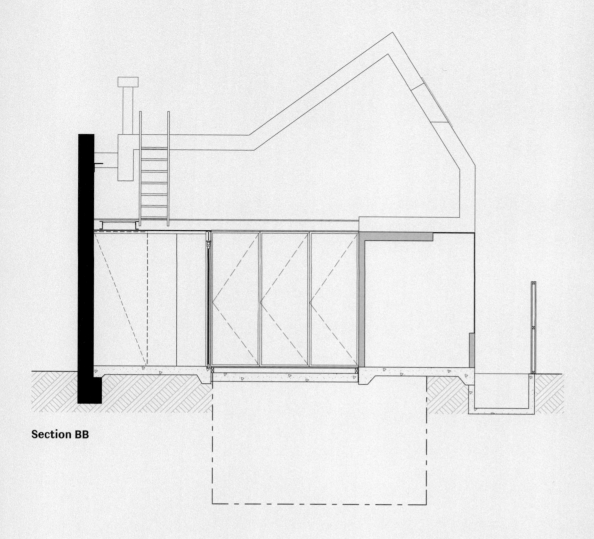

Section BB

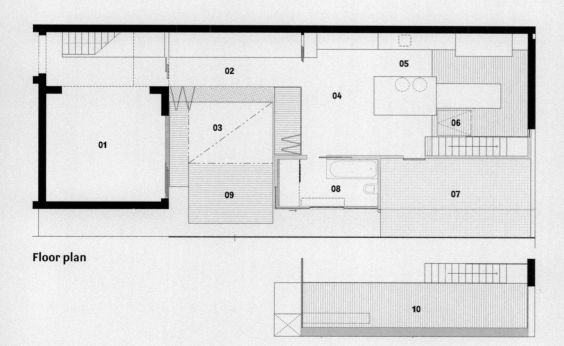

Floor plan

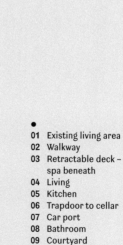

●
01 Existing living area
02 Walkway
03 Retractable deck –
    spa beneath
04 Living
05 Kitchen
06 Trapdoor to cellar
07 Car port
08 Bathroom
09 Courtyard
10 Mezzanine level

Black Line One X Architecture Studio
PROFILE HOUSE
Brunswick East, Melbourne, Australia, 2012

An iconic addition to a quaint bungalow offers an evocative tribute to the defined industrial typology found in a northern suburb of Melbourne. Highlighting rather than erasing its historic roots, the extant house daringly preserves its original face. The harmonious design instead introduces a graphic, 2D treatment along its western elevation that visually unites the old and new parts of the house. Clad with rough sawn Victorian Ash timber, a strong internal vertical profile extrudes to the exterior to form a distinct silhouette toward the street. Transitioning from fence to façade to door, the articulated wooden surface represents the defining architectural expression of the project while creating a more secure perimeter to enjoy a private family life.

A refreshing change from housing that simply maximizes developer returns, the low scale redesign maintains the area's built integrity. Located in a semi-industrial zone, the site offers visual proximity to the neighborhood's brick and corrugated iron warehouses. With some warehouses and factories still operational, others lie dormant and in disrepair. Developing an engaging interplay between past and present, the project builds on this rich existing context and site parameters. In a subtle move, the timber boundary walls soften the corner and edges of the site.

THE LIGHT WOOD EXTERIOR PLAYFULLY OSCILLATES BETWEEN FENCE, BARN, FUN HOUSE, AND WALL, AS ITS IRREGULAR ROOFLINE PEAKS AND DIPS. Meeting the low brickwork framing the front of the white, craftsman façade, the wooden fence begins its ascent as it steps up to shield the side of the original house. The fence again steps up in height to extend

over the entire surface of the home's addition at the back of the lot. After reaching this dramatic height, the wood siding dips down to once again return to its fence typology. This evolution, from fence to wall to fence again, enables this part of the structure to double as a hidden door into the backyard. By ending on a modest note, the back part of the fence terminates at the exact same height as the neighboring brick garage built right against it. A sole square window permeates this wooden façade, adding intrigue as people pass by, and relating back to the language of the house. The charming language of the wooden intervention references the classic warehouse silhouette, with its tilted roofline for clerestory windows. This addition whimsically juxtaposes with the materials and utilitarian nature of the surrounding factories. Reappearing brick elements threaded through the initial structure further ground the project to its context. Bridging the gap between the different parts of the house and the different buildings on the street, the exterior update finds a sensitive yet visionary approach to the future.

THE OPEN AND FLEXIBLE INTERIOR SEAMLESSLY WEAVES THE ADDITION BACK INTO THE ORIGINAL AREAS OF THE HOUSE. High angled ceilings stand as the defining character of the continuous internal spaces. The main living areas open up to a garden and a courtyard of wharfdecking, establishing a fluid transition between indoor and outdoor living. A smaller lightwell to the west allows natural light to penetrate throughout the entire plan. In addition to the retention of three front rooms and a central bathroom from the extant house, the plan now features a reworked kitchen and dining area, an informal children's area, and a vibrant living space. A new master bedroom, en-suite, and private garden are also incorporated. These imaginative and interconnected spaces result in a home that feels much larger than its footprint. The overall design strengthens family engagement and visual communication, reinforcing the relationships between space and activity.

Simple planning creates clean, elegant, efficient, and sculptural internal living spaces. According to the clients' wish for a high-performance sustainable house, their desired spaces evoke a feeling of openness, lightness, and visual continuity that connects the home to the landscape. These spatial values are reinforced by the clients' professions, with one specializing in environmental site characterization and remediation, and the other practicing alternative medicine. The dependable design strategies at play provide careful plan orientation and cross ventilation. Other sustainable elements include solar hot water and water harvesting systems as well as the use of north facing thermal mass in the main living space. Throughout the addition, natural material choices and finishes, including unseasoned, unmilled Victorian ash timber and bamboo flooring, adhere to the clients' environmental stipulations.

This compact, economical, and highly livable outcome manifests as the result of an inclusive dialogue between all consultants, the builder, the architect, and most importantly, the clients. Encouraging the architectural elements to disappear into the background, the design takes the archetypal bungalow and lightens it while remaining responsive to its original vernacular and detailing. The house works with rather than against existing forms. Carrying the themes from the old structure into the new, the residence embraces the continuity of change as it evolves from a traditional house into a contemporary home. ■

PLANS & DRAWINGS

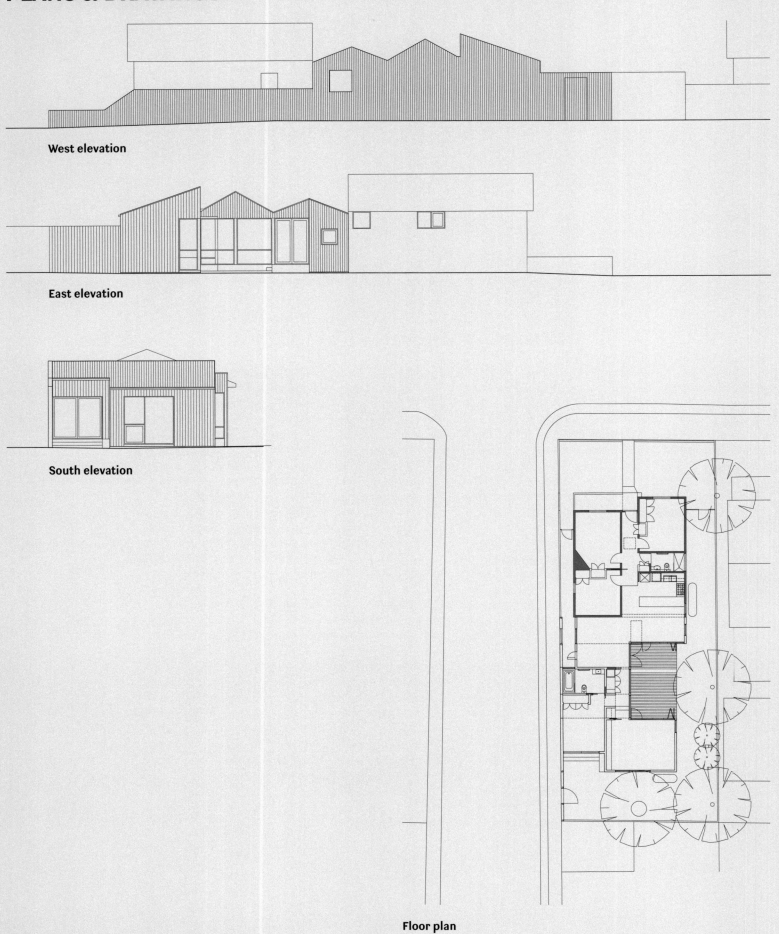

West elevation

East elevation

South elevation

Floor plan

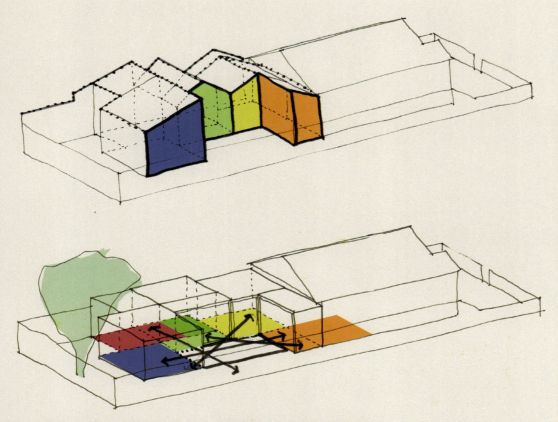

Diagrams showing the interconnected nature of the new addition

Spatial Strategy

This long corner plot enables an original craftsman to expand down the street. A new set of bright and modern living spaces opens up onto an outdoor terrace in the backyard to form a hybrid inside/outside space.

●

Number of square meters:
170
Number of rooms:
10
Number of floors:
1
Number of residents:
3 adults, 2 children
Sustainable features:
Cross ventilation,
solar hot water,
water harvesting system,
thermal mass via a concrete slab,
natural, untreated, and sometimes
recycled external materials and
finishes with low embodied energy,
bamboo internal flooring,
double-glazed windows
Outdoor areas/rooftop access:
Deck and garden
Context:
Deck and garden
Type:
Addition and renovation

Section 1

Section 2

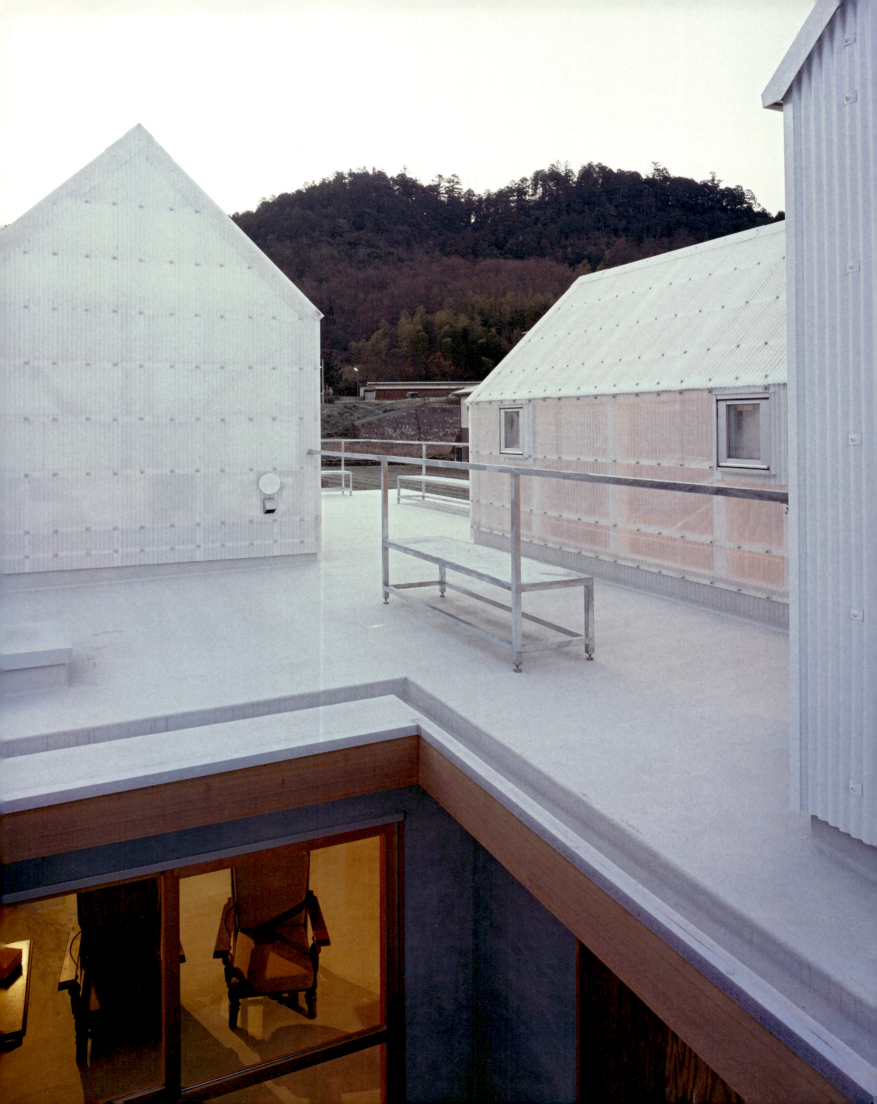

Tato Architects
HOUSE IN YAMASAKI
Hyogo, Japan, 2012

■ A house for a couple and their two children sits in a neighborhood surrounded by mountains and overcast skies. With the goal of letting in as much daylight as possible, three archetypal house volumes, varying in scale and transparency, rise up from a 1.8-meter-high grey platform. Evoking the density of a micro city, these pitched roof structures accommodate a guest room, bathroom, and sunroom. The formal language of the project inverts the iconography of the house, as the three white volumes function more as climactic tools for the spaces located on the plinth below than as spaces themselves. This reversal of occupiable areas allows the family to keep a distance from neighbors and passersby while maximizing open space for the children to play.

In addition to collecting heat in the winter, the sunroom doubles as a greenhouse and brings in fresh air through five motor-operated windows. The translucent bathroom and sunroom glow brightly during the day and transfer sunlight and ventilation to the ground floor via strategic openings and circulation passages. A layer of insulation between the polycarbonate walls of the bathroom prevents this private space from becoming too transparent and exposing more than just a ghosted silhouette. Introducing an additional level of privacy, the opaque walls of the guest bedroom ensure a discreet atmosphere for relaxation. Rooftop paths, complete with built-in benches, enable easy access to the distinct volumes on the second level. A scattering of potted trees and plants stand out against the pristine white backdrops of the three sheds. These outdoor areas also double as overhead terraces and

lookout points, offering spatial and visual connection to the ground floor rooms as well as the city beyond.

The first floor, partly sunken below ground level, responds to a challenging sloping site while using the earth as a natural form of insulation for floor heating.

This lowered level holds the main living spaces of the house as well as the two bedrooms and kitchen. Simple wooden planes break up the open layout to add a level of intimacy to the sleeping areas. These partitions, reminiscent of large shipping containers, also demarcate space for storage and a smaller bathroom. A long desk runs across one side of the

Located in a new residential area bordered by fields, the house reads as a hybrid cluster of agricultural huts.

main living space. Two sets of overhead shelving complete this shared work space and border a window that looks out onto the neighborhood. A large picture window orients the living room toward an outdoor courtyard. This tranquil space, referred to as an outer room, further embeds the outdoors in the inner workings and rituals of residential life. **BY BRINGING THE ROOFTOP CLOSER TO THE GROUND AND SUPPORTING SPACE FOR AN OPEN-AIR SANCTUARY,**

THE ENTIRE SITE BEHAVES AS A TYPE OF CONTINUAL GARDEN WHILE MAINTAINING BROAD VISTAS OF BOTH MOUNTAIN AND SKY.

Located in a new residential area bordered by fields, the house reads as a hybrid cluster of agricultural huts. Polycarbonate corrugated panels, a vernacular material in the region, clad two of the three independent rooftop volumes. This rustic cladding relates to the existing agrarian landscape and provides insulation against the cold and moisture buildup. Outer walls of the foundation platform are covered with fiber-reinforced cement boards that leave space for rainwater runoff and add shaded contrast against the façade.

The imaginative layout strikes a satisfying balance between inward and outward reflection. Connected via ladders and a white staircase, the relationship between the two levels of the house capture the imagination—a whimsical treehouse for all ages. Unexpected juxtapositions occur

between the top and bottom floors. From the modest dining table, family members can look up through the main open light well to observe who is studying diligently at the table that runs along the edge of this void. After being called down for dinner, the stray parts of the family can quickly climb down the ladder and rejoin at the table. Placed squarely under the cutout to the upper level, this dining area enjoys ample illumination and an enhanced feeling of spaciousness as the double height area floods with light. The opening between first and second floors can be closed with a shade during extremely hot or cold times of year to maintain a stable indoor climate regardless of the season.

By day, the house fills with the splendor of diffused sunlight that intermingles with its crisp, white interiors. By night, the most private lower level of the home casts a warm glow into the community through large, square windows. Framing views across spaces and across the land, this uplifting design creates a sense of programmatic freedom. Accented with light wood furniture, sheer mandarin orange curtain partitions, and the occasional lemon yellow surface, the airy and interlocking rooms evoke a type of Japanese bricolage for everyday living. Minimal without being sparse, the adventurous residence sets up a harmonious relationship between nature, the city, and the family unit. ∎

PLANS & DRAWINGS

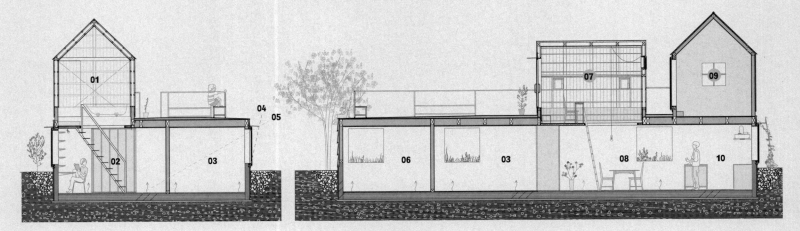

Transverse section

Longitudinal section

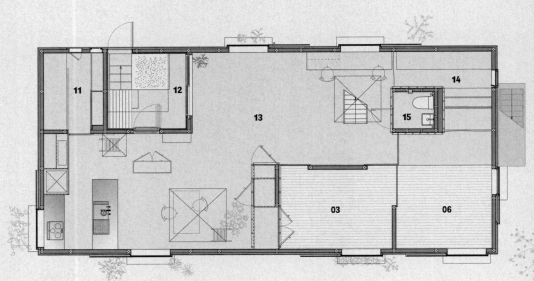

Ground floor plan

- 01 Bathroom
- 02 Living
- 03 Room 1
- 04 The summer solstice
- 05 The winter solstice
- 06 Room 2
- 07 Sunroom
- 08 Dining
- 09 Room 3
- 10 Kitchen
- 11 Closet 1
- 12 Outer room
- 13 LDK
- 14 Closet 2
- 15 Half bath

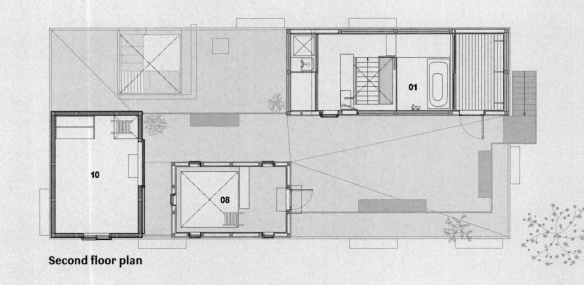

Second floor plan

Spatial Strategy

Three translucent and independent rooftop structures direct light into the residential plinth below. Forging numerous connections between levels, the bright open spaces remain connected to the outdoors via a central courtyard on the lower level and a meandering rooftop terrace above.

●

Number of square meters:
119.11
Number of rooms:
6
Number of floors:
2
Number of residents:
4
Sustainable features:
Thermal insulation
Outdoor areas/rooftop access:
Courtyard and rooftop terrace
Context:
Outskirts
Type:
New construction

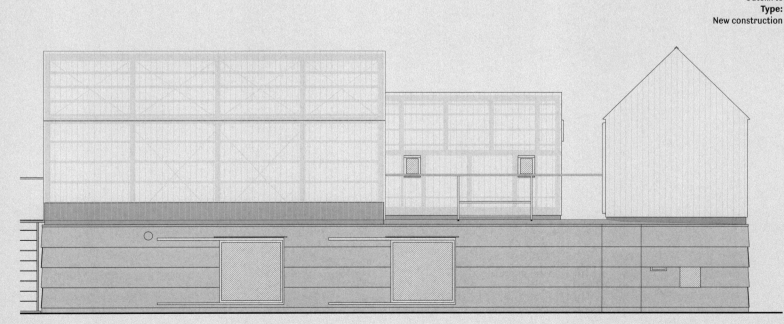

East elevation

North elevation

<u>House on Mountainside Overlooked by Castle</u>

Fran Silvestre Arquitectos
HOUSE ON MOUNTAINSIDE OVERLOOKED BY CASTLE
Valencia, Spain, 2010

■ On a mountain topped by a castle, this striking white residence capitalizes on the unique beauty of the landscape. The site, set int the outskirts of town, stands as the result of both natural and evident growth. The structures found in this area adapt to the distinct topography. However, these structures merely act as blanket housing—a system of aggregated and fragmented pieces. Within this more uniform architectural context, the new building behaves as a dramatic introduction to the contemporary aesthetic. Nestled into the rock, the striking geometric formation of the house commands attention.

From the street, the home's futuristic main façade activates a series of tilted planes. As some fold in, others fold out to create overhangs for upper level balconies. From these protected, angular perches, the residents can take in the sweeping panoramic views of the city. Unadorned sheets of white introduce a new architectural agenda into an area filled with an old world vernacular based on intricacy and a warm material palette. By comparison, the sharp, smooth walls of the house shake off the ornamental nature of the past. Even so, the project integrates into the environment, respecting the strategies of its neighbors while still crafting a unique identity in terms of both style and material. Shying away from misleading historicism, the design catapults the neighborhood into the future. The residence

delicately fills in the gap of the land. Enhancing the feeling of lightness, the gleaming, all-white building seems delicately placed on the ground.

Divided by a large light well, the indoor space arranges a parallel interior environment next to the mountain. The ground floor hosts the garage and cellar. Resting on top, another volume consists of two floors with four rooms. Two of these rooms are

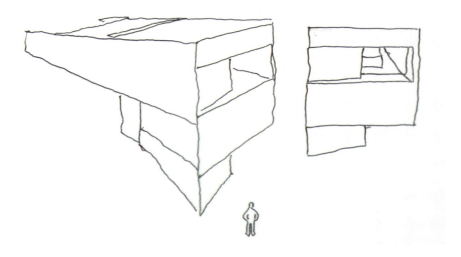

located at the intermediate level and open up to the private street outside. The other two spaces, located on the upper level, overlook the surrounding houses

FORMAT, THIS TRANSPARENT THRESHOLD ENHANCES THE VISUAL EFFECTS BETWEEN INTERIOR AND EXTERIOR SPACES. Adding to this feeling of contrast, the matte gray materiality of the terrace stops precisely at the line where the kitchen starts. From here, a universe of glossy white surfaces begins. A sleek modern kitchen rises in the center of this adjacent interior.

An open kitchen features a low rectangular kitchen island backed by a prominent wall with a glowing cutout for cabinetry and appliances. Reaching the ceiling, the back wall of this kitchen area doubles as

and the valley of Ayora. One of these rooms, the study, opens up to the central double-height space at the core of the home. By forging this spatial connection, the study increases in grandeur and unspoken eloquence. The areas facing the garden and mountain appear across this gap. These spaces bask in the natural sunlight that reflects on the southern slope of the oxidized castle.

From the back, the mountain levels out to form a rugged path where only the top floor remains visible. This top level

reads as a monolithic white rectangle, cantilevering into the great beyond. Perceived as a complete and solid box, the walls of the plinth successfully hide an outdoor terrace nested just within. The private terrace basks in the exhilarating views of the towering mountains that pick up where the house leaves off. **REACHING FOR THE SKY, THESE TOWERING FORMS SET UP A DYNAMIC CANVAS WHERE THE MANMADE AND NATURAL WORLDS INTERSECT.** A single door, camouflaged into this outer wall, connects the house to its rustic backyard—the mountain. The terrace, bordered on three sides by the high concrete walls, fluidly merges inside with out. Establishing contrast with both the perimeter exterior walls and the brilliant interior spaces, this terrace offers a space of contrast by keeping the dark gray tones of its exposed concrete surfaces intact.

The final wall of the outdoor space comprises floor-to-ceiling glazing with doors that can swing open on either side. **APPLYING A DRAMATIC RECTANGULAR**

a method of demarcating one space from another. In this instance, the wall in question becomes a void on the other side. Perhaps the home's most mesmerizing feature, this central void operates as an ethereal atrium space. The multilevel core frames views of the bright blue Spanish sky. Evoking some of the best artworks of James Turrell, the generous skylight above floods the interior with light, peacefully marking the passage of time. Defying expectations for how a residential project can feel, the unusual and uplifting space of intersecting planes oscillates between the muscular, and highly sculpted, and all things ephemeral and unbelievably light.

A driveway cut into the rock leads to a hidden entry halfway up the side of the house. Tucked back into the peripheral wall, the main entrance achieves a rare sense of privacy away from the busy street. Warm light flows out of this entry point and nearby windows in the evening, transforming the structure into a welcoming and intriguing beacon for the neighborhood. ∎

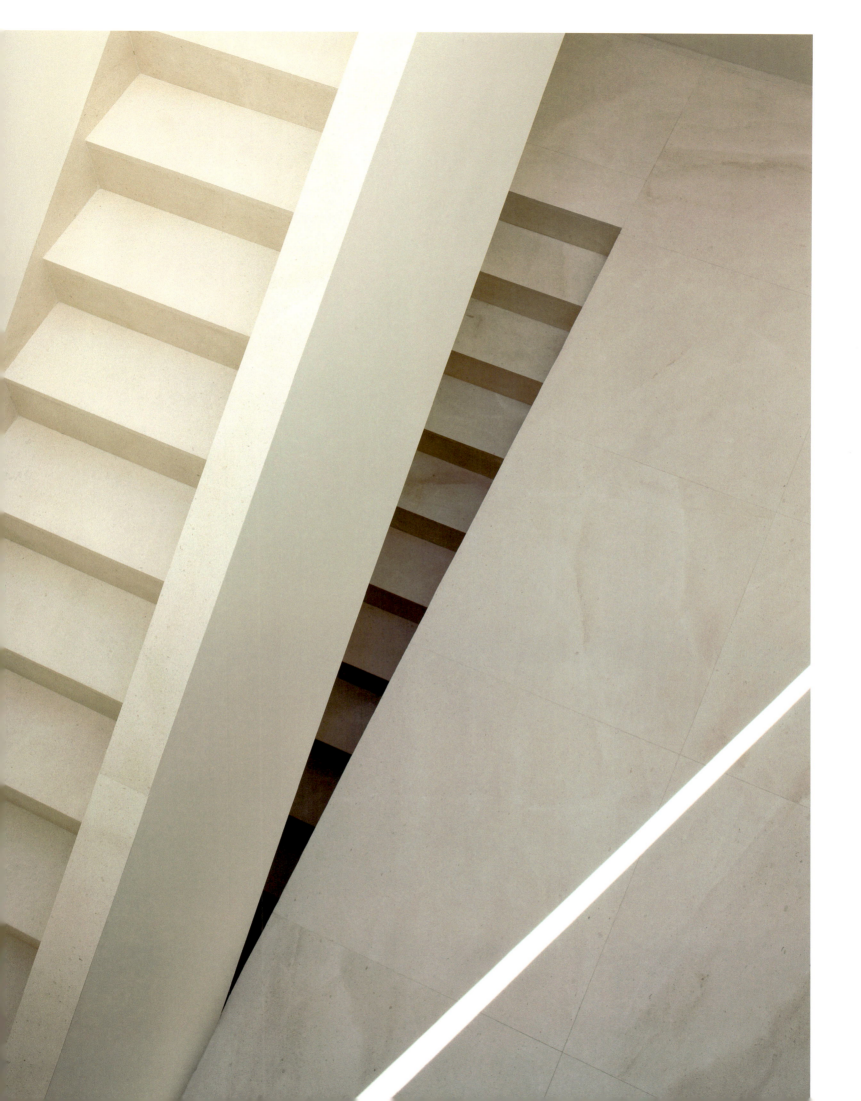

PLANS & DRAWINGS

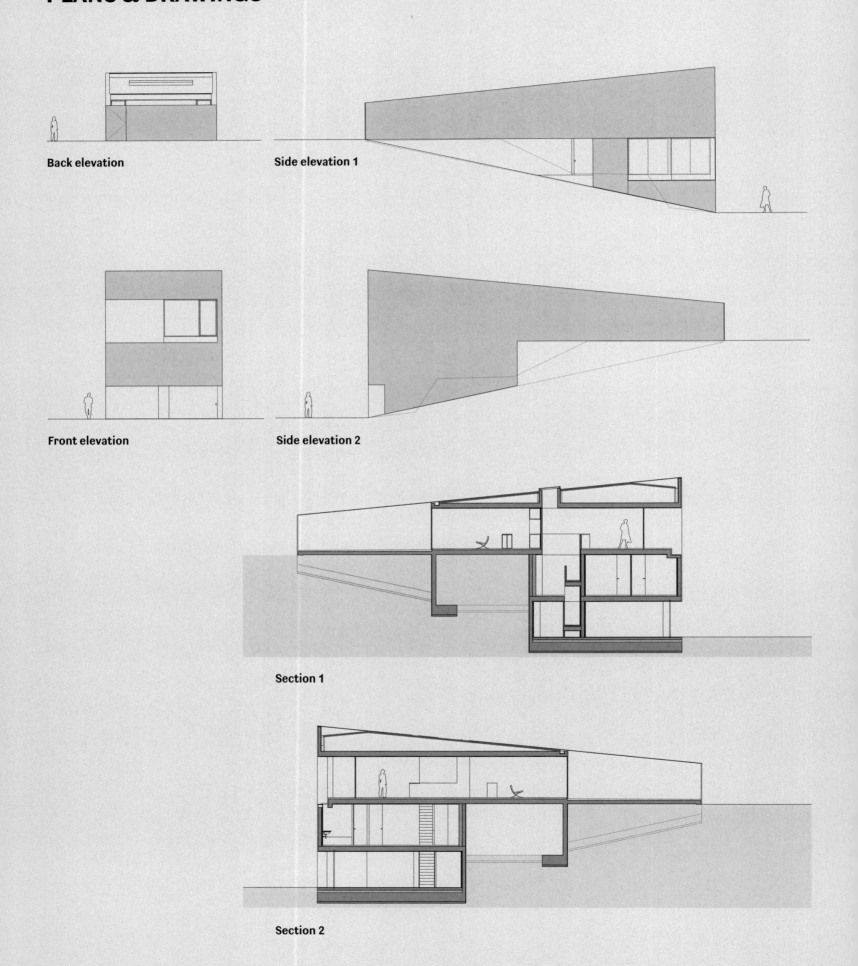

Back elevation

Side elevation 1

Front elevation

Side elevation 2

Section 1

Section 2

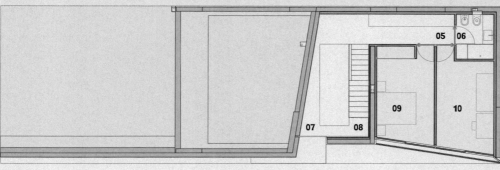

Ground floor plan

HOUSE ON MOUNTAINSIDE OVERLOOKED BY CASTLE

Spatial Strategy

The interior spaces of this three-level house wind around a bright atrium. Nesting into the rocks behind, the toplevel connects out onto a private rooftop terrace.

●

Number of square meters:
230
Number of rooms:
10
Number of floors:
3
Number of residents:
4
Sustainable features:
Efficient insulation
Outdoor areas/rooftop access:
Terrace
Context:
Outskirts
Type:
New construction

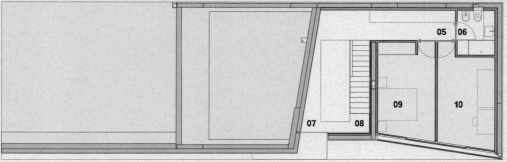

First floor plan

●

01	Wine-cellar/store room
02	Hall
03	Garage
04	Laundry room
05	Distribution board
06	Bathroom 1
07	Reception
08	Stairs
09	Bedroom 1
10	Bedroom 2
11	Terrace
12	Living/dining
13	Kitchen
14	Studio
15	Bathroom 2
16	Bedroom 3

Second floor plan

HOUSE IN KEYAKI

Honjyo, Japan, 2012

■ This bronze-colored home boldly commands a corner lot an hour and a half drive from Tokyo. Surrounded by neighbors without any getting too close, the unusual site grants the building unobstructed visibility on all sides while remaining integrated within the community context. Due to the lack of public transportation available, the residents of this modest neighborhood rely almost exclusively on their cars when they need to get somewhere. With the car and parking spaces at the forefront of the priority list, the house stands as the result of designing from the outside in.

Metal clad and timber framed, the tilted outline of the three-story residence steps back from the street. This setback makes way for a parking space and a small garden while directing sunlight from all directions into the building. Protected by a simple overhang, the understated front door blends into the façade by applying the same metallic cladding. The humble entry sequence still references the youthful nature of the family inside with a single concrete step that transitions into a micro skateboard ramp. However, this subtle detail serves as one of the only clues about what can be expected from the interior—a lofty, handmade space with a treehouse atmosphere.

An all but invisible roof featuring the unique bronze siding dissolves into the overall exterior wrapper for the house. The thinness of the roofline allows the residence to read more like a compelling sculptural object than a house from the outside. Adding to the found object quality of the home, a slender gap runs along the perimeter of the exterior, keeping the façade just a few centimeters away from the foundation. **THIS INCONSPICUOUS SHADOW GAP GIVES THE BUILDING AN EXTREMELY LIGHT READING, SEEMINGLY HOVERING ABOVE THE GROUND.** The visual surrealism present across the exterior continues with the discreet manipulation of scales occurring on the

main façade. Angling sharply up, the house gains height as it reaches the street corner. Here, this tallest point of the home corresponds with the location of its smallest door. Instead of capitalizing on the gained height to make way for a prominent entry, the front door to the house instead feels miniature-sized

in comparison to the grand scales that surround it. The same technique appears again on the smallest part of the house, with the introduction of the tallest window and sliding door system. Such inverted relationships and plays on scale keep the house in a state of dynamism where the eye never settles.

WHILE CERTAIN FACES OF THE HOME REMAIN COMPLETELY CLOSED OFF TO THE OUTSIDE WORLD, OTHER SIDES CAREFULLY FRAME VIEWS OF NATURE, CITY, AND SKY. Three tall, narrow windows perforate the south, east, and west sides of the house. The light entering from these openings reflects onto the white walls, wrapping the objects and people inside. The thoughtful location of these windows tracks the sun

while demarcating the transition between rooms. Continuing with this theme of flexibility, rotating white panels act as internal windows that can open up or close off certain rooms from one another. These easily operable panels also help transfer daylight into the darker areas of the house.

On the ground floor, the placement of the stairs and a shift from wood to concrete flooring separates the open kitchen and living room. The same staircase breaks up the main bedroom on the second floor into places for sleep and study. A top-level mezzanine area serves as a second living room that opens out onto a rooftop terrace. The outdoor terrace behaves as a cutout—a shallow rectangle removed from the slope of the roof. By keeping the lines of the roof intact, the stealth void proves almost invisible when viewed from the street. Only when occupied does the presence of the terrace become clear, as the people engaging the space playfully pop out above the shallow retaining walls. The angled profile of this outdoor area grants it a dynamic quality that propels guests right out onto the edge of the city. From this terrace, one can slip away from the worries of city life and revel in the majestic mountain views of Northern Japan. This enchanting project contrasts scales, shapes, and materials to elicit a memorable experience from both inside and out. ∎

throughout the day and its changes from season to season. An interior setback in front of the windows on the second floor brings light deep into the house. These voids also provide visual connections between the outside and inside of the building, allowing its occupants to anticipate the presence of one another across the different rooms. This overlapping of vantage points and nested spaces results in a floor plan that feels much larger than it is. Wood floors and simple lamps dangling from the exposed ceiling rafters lend the spaces a warm and endearing quality. Strategically placed furniture and built-in shelving form borders and subtle partitions between the triangular light wells and the surrounding living areas.

Divided into a series of small spaces, the flexible layout can be adjusted to the residents' preferences. Wooden shelving and a central stairs add privacy

PLANS & DRAWINGS

First floor plan

Second floor plan

Roof plan

-
01 Pantry
02 Bathroom
03 Light well
04 Living
05 Kitchen
06 Dining
07 Entrance
08 Room 3
09 Half bath
10 Room 2
11 Room 1
12 Rooftop
13 Loft

HOUSE IN KEYAKI

Spatial Strategy

Accommodating a parking spot onto the narrow site, this commuter house generates open connections across three levels. Strategically placed windows allow light and glimpses of nature to infiltrate the interior, while a semi-hidden rooftop terrace becomes the prize lookout point for the residence.

Number of square meters:
103.13
Number of rooms:
3
Number of floors:
3
Number of residents:
2 adults, 1 child, 1 dog
Sustainable features:
Natural Ventilation
Outdoor areas/rooftop access:
Rooftop terrace
Context:
Residential suburbs
Type:
New construction

Section 1

Section 2

02 Bathroom
04 Living
06 Dining
07 Entrance
08 Room 3
10 Room 2
11 Room 1
12 Rooftop
13 Loft

Areal Architecten
HOUSE DV WILRIJK
Wilrijk, Belgium, 2012

■   This contemporary home stands as the final accent at the end of a street lined with typical Flemish row houses. The dark sculptural volume overcomes the constraints of the traditional housing typology while remaining careful not to overshadow its rich legacy. Aligning perfectly with its historical neighbors, the systematic façade with scattered openings wraps a refined interior composed of spit-level floors.

Thanks to a difference in ground level with the neighboring residence, the home office nestles into a semi-basement. This partially lowered floor gradually turns into a living room. Due to the principal floor's central placement, this level releases a range

of viewpoints and internal relationships between the different spaces. Despite its slightly inclined roof, the house follows the guidelines set in place by this urban settlement and challenges the region's classic vernacular. The metal plated light-weight

skin of the façade gives the volume a certain autonomy. **RESEMBLING A BLOCK OF BASALT FORMED UNDER THE INFLUENCE OF ITS ENVIRONMENT, THE STEALTH FAÇADE WALKS A FINE LINE BETWEEN BLENDING IN AND STANDING OUT.**

Due to budget limitations, the sustainable features are reduced to passive measures including natural ventilation. The house is heated by the combination of a low-temperature floor-heating system and radiators. Air-tight details between walls and windows minimize drafts, promoting efficient heating and cooling. Executed on an impressively modest budget of 215,000 €, the budget directly influenced the design approach. The economical project avoids the need for an expensive foundation by instead applying a light façade material that requires only a moderate amount of support. In spite of the thin façade, the design still applies a substantial insulation layer from within. Cast on site, floor slabs made from wood plank formwork lend the home a slightly rugged flair. This added detail enhances the atmosphere and material texture of the house.

Built for a family of three, the three-bedroom house consists of a total of 12 functional spaces. The pitched roof residence, set on the street's corner plot, maximizes light, space, and privacy. Respectful yet undeniably unique, the home connects to the language of the street while maintaining a strongly introverted interior. After accounting for the home's striking monolithic color and nature, the next differentiating characteristic appears in the location of the front door. The house takes advantage of it corner location and instead orients its main entrance along the side. By differing from its neighbors and placing

the main entry on the side street, the house shields the comings and goings of its guests from the public eye. The act of making the corner the front supports the impression that the house is as public as its row house neighbors, but in fact it proves far more private. Even so, this move toward privacy still activates two of the building's faces along the street.

What at first glance seems to be the front façade, in fact reveals an assortment of windows of varying sizes. Beginning at ground level, an off-centered, narrow rectangular window brings light into the basement and entry areas. A significantly larger window appears a level up, framing views of the outdoors for the living room. A smaller square window towards the top and a similarly sized skylight complete this front façade.

THE FACE OF THE MAIN ENTRY ALONG THE SIDE GROWS IN COMPLEXITY. Showcasing an array of square windows, these recessed openings protect from the rain. Two larger setbacks integrate the front door and a modest balcony above. An angular paved walkway leads up to the entry from the street. This entry sequence pushes the walls of the house inward to construct a protective overhang. One level up, a simple pane of glass serves as the guardrail for the balcony. Giving the façade a reflective quality, the understated balcony allows the family to keep an eye on visitors and admire the noble trees lining the opposite side of the driveway.

While the front of the building reads as a flat plane, the back juts out to form a more boxy shape. Here, a large, open backyard defines the composition of the rear part of the house. Large, sliding glass doors link the kitchen and dining areas to the outdoors. A generous paved slab extends these shared interior spaces into the open air for outdoor dining and socializing. With the entire house opening up to the backyard, a contemporary porch comes to life.

CONTRASTING WITH THE DARK AND DRAMATIC EXTERIOR, THE INTERIOR SPACES ENGAGE ALL WHITE WALLS. THESE INTERIOR SPACES RELY ON A MINIMALISM ENHANCED BY NATURAL AND RUSTIC ELEMENTS. The use of a grey weathered wood promotes a more intimate experience of the high ceilings. Cozy, modern, and functional, the interior presents an interconnected network of planes, stairs, and personal nooks. A flight of stairs leads to a loft area dotted with stacks of floating books. This unconventional approach develops a decorative yet easily navigatable solution to the usual bookshelf. Inverting tradition both inside and out, this residence achieves the perfect blend between following the rules and questioning their boundaries. ∎

PLANS & DRAWINGS

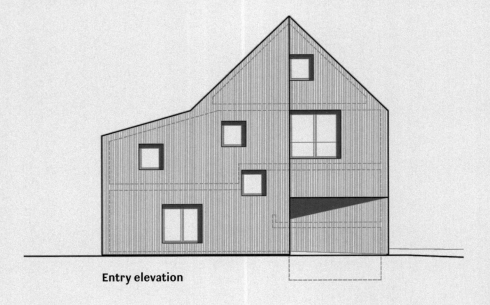

Entry elevation

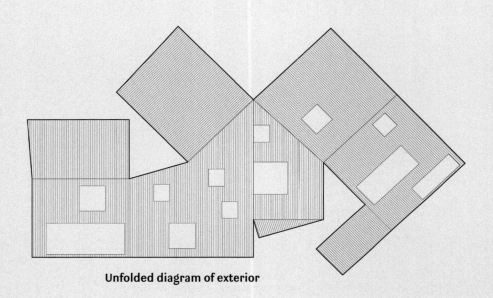

Unfolded diagram of exterior

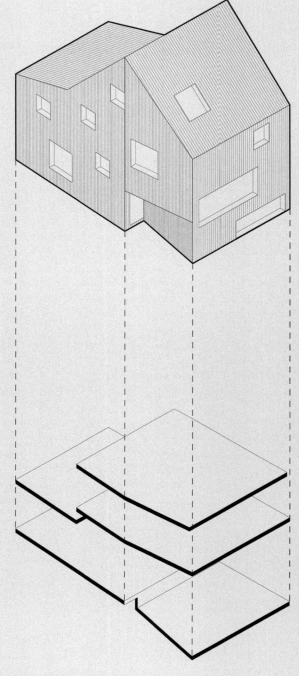

Floor plate diagram

Areal Architecten
HOUSE DV
WILRIJK

Spatial Strategy

A split-level modern home integrates with its more traditional surroundings while still nurturing a unique interior logic. Connected by a series of white stairs, the interlocking spaces move from the more public to the more private the higher up one goes.

•

Number of square meters:
265
Number of rooms:
12
Number of floors:
4
Number of residents:
3
Sustainable features:
Natural ventilation,
floor heating system,
air-tight detailing,
efficient insulation
Outdoor areas/rooftop access:
Backyard and balcony
Context:
Residential area
Type:
New construction

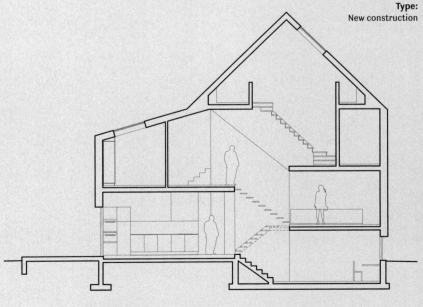

Section 1

Section 2

Lischer Partner Architekten
STADTVILLEN
Lucerne, Switzerland, 2012

150 million years of sedimented rocks, the high quality building material underscores the value of the townhouses. This rare stone also places the homes in close dialogue with the historic structure of the environment. Lightweight walls produce flexible room layouts that respond to the personal needs of the owners and their changing needs over time. Living spaces extend over two floors of detached program. The upper levels host the living and dining rooms as well as the wet rooms and bedrooms.

AS THE RESIDENTIAL UNITS SLOPE DOWN THE HILLSIDE, SO TOO DO THEIR ROOFS. These dynamic core angles generate exciting interior rooms and spatial sequences. Together with the refined material choices and finishes, the well-proportioned rooms and generous ceiling heights set the stage for an energizing home lifestyle in sync with both nature and the city. Translating to "town villas," the name of this residential development succinctly summarizes its typology and general goals.

Responding to the easement guidelines, the multilevel buildings barely protrude above the ground. The four townhouses interact with the city context, taking on the urban pattern of the area. Accented by the spacious villas along the lake, the distinctive townhouses resemble rectangular boulders in a great garden. This low-lying approach cunningly takes a spatial constraint and transforms it into a design asset. **INSTEAD OF REACHING UP, THE HOMES REACH OUT TOWARDS THE WATER.** By taking two horizontal steps down the hill, the layouts open up to the light, air, views, and surrounding gardens.

■   Four city villas, discreetly nestled in the terrain, offer exceptional views and maximum comfort. The site is located on the right bank of the city of Lucerne, well situated in a residential district. A pending redevelopment of the classical homes on the street makes room for these four new townhouses and a below-ground, two-story garage. The site provides an excellent view of the nearby lake and the Alps.

The structure of each building predominantly utilizes reinforced concrete. Solid masonry of yellow Jurassic limestone clads the roofs, while plates of the same material wrap the loggias. Formed out of

Despite their proximity to one another, the homes still serve as immersive private retreats for their residents. This high level of privacy manifests through the careful immersion of each home into the lakeward slope. Such unobtrusive angles help the structures to better integrate with local buildings and streets that define the site.

The houses have no classic windows. As an elegant alternative, skylights and large incisions for terraces and loggias choreograph an enticing interplay between inside and outside. These effective window alternatives bring in plentiful daylight and promote natural ventilation. The surprising interior and exterior relationship reflects the light of the lake below deep into the inner workings of the houses. Generous glass doors extend the living rooms into the greenery outside. This ample glazing fills the interiors with light, defying their proximity to the earth. While hugging the ground, these artful residences embrace the sky.

Sloping white interiors build a momentum towards the outdoors. Framing the breathtaking scenery of the region, the extensive glazing affords the occupants unique and unobstructed vantage points to admire the transition from city to nature. Dark furniture sleekly catches the eye against the backdrop of the bright interiors. Sheer white curtains run across the glazing, encouraging a flexible attitude to the outdoor scenery. Even when closed, these translucent curtains still transmit a crisp, even light into the buildings.

Various green elements equip the outdoor spaces. These natural features include an open meadow bordered by dense bushes and shrubs strategically placed in designed clusters. Surrounding walls of concrete use a special mixture of water and gravel to create smooth surfaces and give the terrain its structure.

Modest overhangs at the front of each building stage covered terraces for the residences. These strips of outdoor space facilitate a lifestyle interwoven with nature. With room for deck chairs or outdoor dining tables, these exterior spaces add a vibrant social heart to the house. During the warm summer months, the sliding glass doors that separate these terraces from the interior living areas open up completely. Once open, the interiors bleed out into the landscape. These humble exterior accents craft exhilarating stages for taking in the iconic views of Switzerland. A large terrace on the second level of each unit introduces a more private space for appreciating the outdoors and enjoying the fresh mountain air. These semi-enclosed terraces link to the upstairs bedrooms. Providing the most private parts of the homes with the most impressive panoramas, the intimate balconies feature floor-to-ceiling glazing that enhances the visceral impact of the setting. Combining themes of prospect and refuge, this efficiently elegant housing development learns from its environment to tailor a bespoke residential experience for its occupants. ∎

PLANS & DRAWINGS

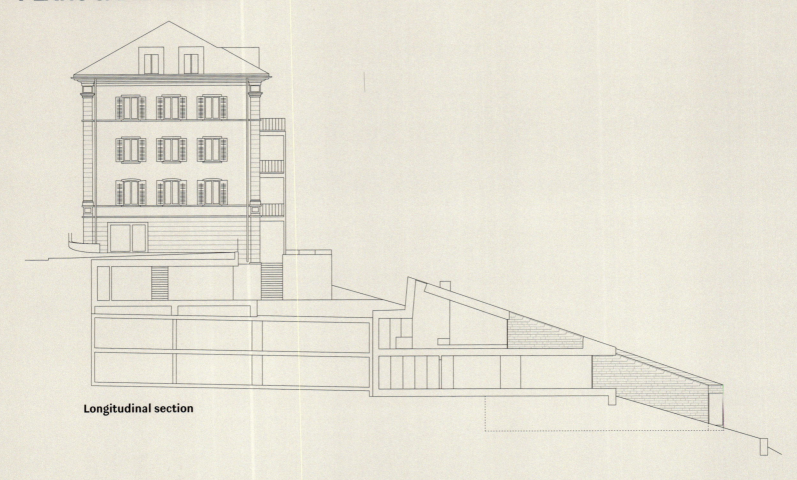

Longitudinal section

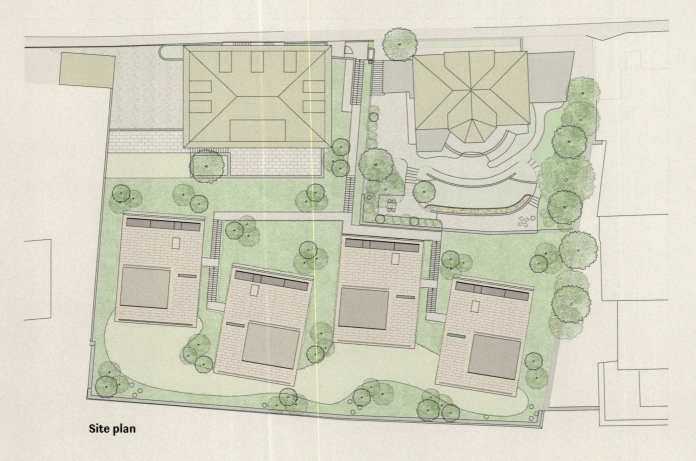

Site plan

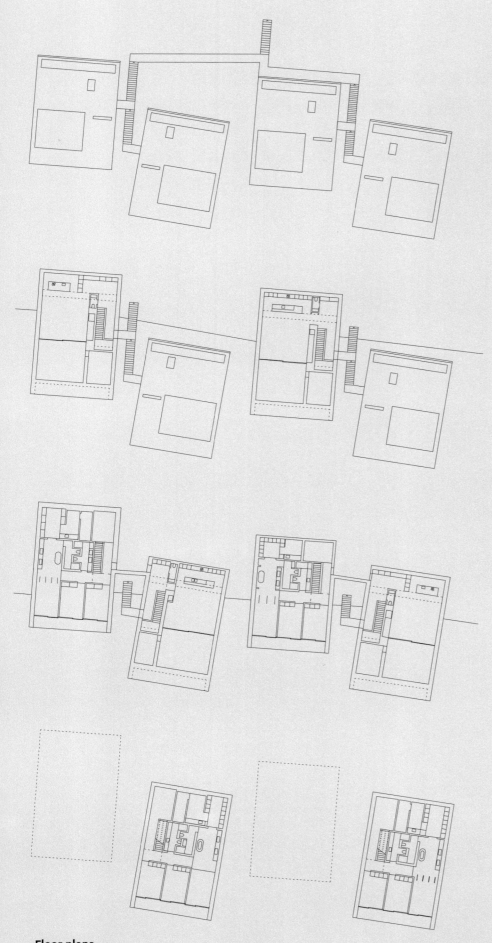

Floor plans

Spatial Strategy

Four individual hillside homes orient toward
a spectacular lake and Alpine view. The
two-level houses arrange the public areas
on the first floor and the private bedrooms
on the second. Rooftop terraces extend the
bedrooms into the fresh mountain air.

●

Number of square meters:
230 per villa
Number of rooms:
5
Number of floors:
2
Number of residents:
Varies
Sustainable features:
N/A
Outdoor areas/rooftop access:
Courtyard and porch
Context:
Hillside
Type:
New construction

Meixner Schlüter Wendt Architekten
RESIDENCE Z
Königstein, Germany, 2012

■   **A**n extension of a 1920s villa, characterized by a roof that slants down on all sides, grows to accommodate a family of five. The upgraded entrance area now integrates storage for bicycles and other useful equipment for outdoor adventures. On the north side of the building, the layout adds a generous living room, bedroom, bathroom, and fitness studio, strategically averted away

from the street. A protected exterior space between both sides of the intervention spreads out in front of the extant part of the house.

Imposed onto the original structure, a thin shell overlaps with the original residence. An array of differently sized windows punctuate both levels of this core part of the house, filling the interior with light. This metallic cladded shell demarcates an enclosed exterior entrance area along its south face. At this entry point, the shell adapts to the traditional angled roof outline of the existing building. As if cut from a larger form, the slate grey shell is exposed to reveal a crisp white undersite. Following this high-contrast reveal, the roofline then lowers further down to protect against solar gains from the western sun. The gleaming skin simultaneously overhangs on the north side, introducing more volume to the newly completed interior spaces within.

Where the transition to the expanded interior begins, the roof folds into irregular triangular-shaped geometries. Generated out of functional considerations, the specific and individual nuances of the roof allow it to evolve from an archetypal pitched roof silhouette to a modern cubic structure. This transformation of the roof culminates in a three-story residential tower that looks out over the backyard swimming pool. Here, a hovering interior concrete volume protrudes through the glass façade onto the exterior to form a balcony for the floor above. While very expressive on the inside, once these volumes reach the exterior they align with the strict grid of the glazing. This compact façade succeeds in balancing weight with transparency. Promoting strong roots to the outdoors, the tall and slender extension offers floor-to-ceiling glazing, balconies, and loggias on every floor. The ground level of this volume orients a spacious living room toward the garden, encouraging the family to enjoy unobstructed views of the backyard and surrounding nature.

ACTING AS A HOMOGENEOUS SKIN, THE ROOF AND THE WALL SURFACES ARE UNIFORMLY CAST IN THE SAME MATERIAL. Referencing the traditionally

This transformation of the roof culminates in a three-story residential tower that looks out over the backyard swimming pool.

shingle-clad roofscapes found in the villas of the neighborhood, the façade also features shingle cladding. The characteristic slateshingles of the region find a contemporary reinterpretation in both scale and material. Wrapped in dark, weathered copper sheets, the shell resembles an enlarged shingle structure. Spanning both new and old areas of the residence, this unusual folded roof serves as a feat

to the current high efficiency standards. Other sustainable features include the planting of new trees and bushes as well as the re-cultivation of a formerly canalized stream that runs through the property.

THE 13-ROOM HOUSE SITS ON A SPACIOUS SITE SURROUNDED BY OLD TREES. Wooden terraces line the east and west sides of the residence, promoting a

of engineering and complex geometry management. Such involved detailing requires a high degree of flexibility and persistence on the part of the structural engineer and the construction crew.

While the existing section of the house stands as a fixed element that defines this part of the new shell from within, the added living spaces behave as inscribed volumes embedded into the exterior canopy. The goal of connecting the old and the new into something coherent represents the biggest challenge facing the design. This all-encompassing shell not only unifies the original house and its extensions into a singular, modern gesture, but also generates an improved weather and thermal protection layer. The design and inherent logic of its updated form dramatically improves the old home's energy consumption. Meeting the German regulations for energy saving buildings and building systems (EnEV), the thermal protection present throughout the house correlates

lively relationship to the outdoors. Opening up to the lush garden beyond, the pristine pool sits adjacent to the main western terrace.

Although the exterior focuses on the unification of disparate elements, the interior still retains distinctive qualities that demarcate the original areas from the newer spaces. The inside of the original house keeps its classic language. Exemplified by the library's rhythmic wood flooring, built-in book shelves, and molded fireplace, these time-tested areas of the house still find new ways to relate to the lush outdoor scenery. In this instance, the centrally located library extends out onto the backyard deck via gracious sliding doors. The visceral juxtaposition between the dark textured shell and the clear, light, and often fully glazed interior areas sets up a dynamic threshold between inside and out. ■

PLANS & DRAWINGS

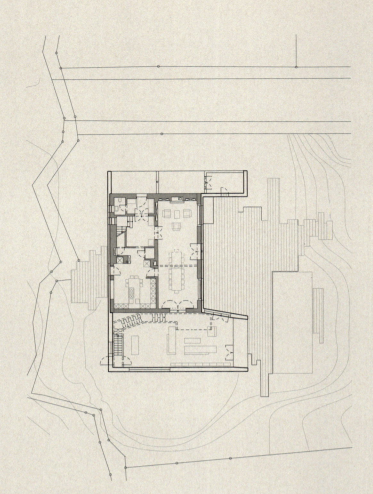

Ground floor plan

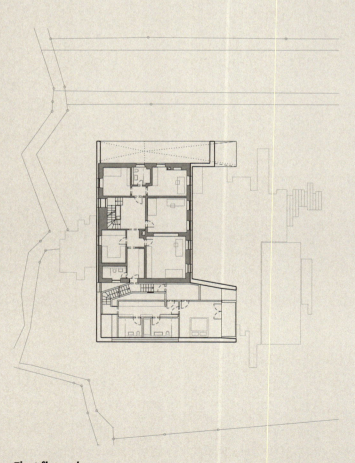

First floor plan

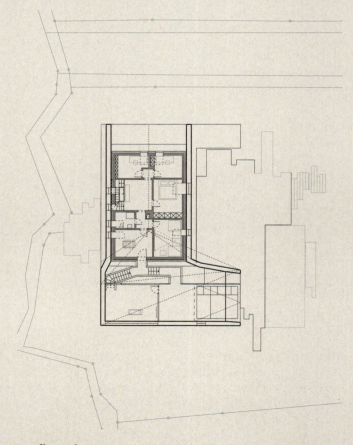

Top floor plan

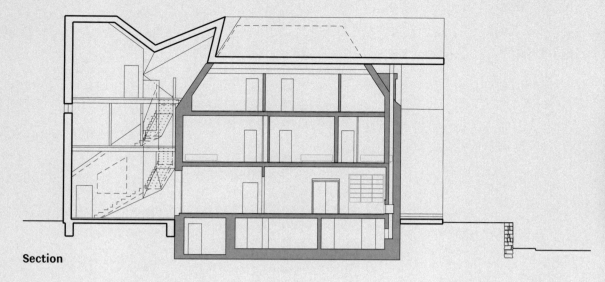

Section

RESIDENCE Z

Spatial Strategy

A metallic pitched roof unites a renovation of an existing house with its modern extension. Looking out over a pool surrounded by a deck and garden, the three-story extension includes open living areas on the bottom and more intimate bedroom spaces on top.

Number of square meters:
557
Number of rooms:
13
Number of floors:
3
Number of residents:
5
Sustainable features:
Thermal protection layer
Outdoor areas/rooftop access:
Terraces, balconies, swimming pool, and garden
Context:
Outskirts
Type:
Addition and renovation

Spatial extension

Entrance/ enclosure

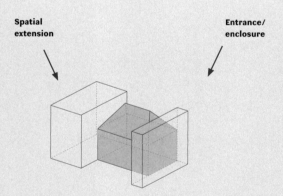

1   Existing building and extension volume

View/garden outlook

Entrance

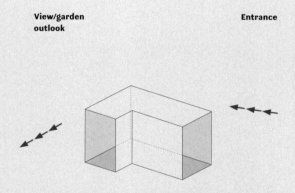

2   Cladding: spatial layout, entrance, and outlook

View/garden outlook

Entrance

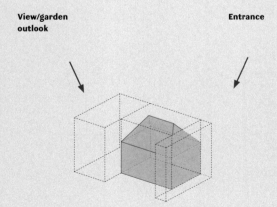

3   Transformation through the existing building and the extension

Transition from gable roof to cube

Roof slope for sun

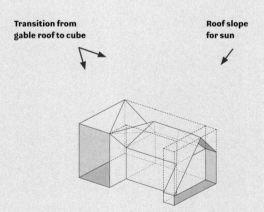

4   Adaptation to the existing building

Jonathan Tuckey Design
SHADOW HOUSE
Colerne, UK, 2007

■ With imagination and know-how, a Baptist chapel takes on its second life as a single-family home. The generous and welcoming house seamlessly preserves the historic character of the original structure while adapting to meet the needs of a young family. Keeping the charming chapel entirely intact in the front, a new extension at the rear stealthily links old and new areas via a transparent glazed passage. This transitional area echoes the vernacular tabernacle churches of the region while softening the transition between the two buildings. Complimenting the chapel's form and scale, the new extension clad in blackened larch wood remains unseen from the street—a quiet shadow of the original building.

Built in 1867, the elegantly restrained Providence Chapel features creamy Bath stone. The dignified, symmetrical elevation, with its generous arched windows and simple Roman Doric porch, addresses the center of the village. Set within a compact plot defined by dry stone walls, the ecclesiastical building retains its unadorned interiors and vernacular detailing. The vestry building now acts as the home's library. Its large, rustically converted hall accommodates an open kitchen, living room, and mezzanine gallery above accessed via two sets of stairs located on both sides of the entry. The lofty double-height room embraces its historical past. Mismatched chairs set around the dining table, original wood flooring, endearingly weathered rugs, and a centrally placed antique clock on the mezzanine all influence the mixed-use space's informal quality. Coated in a slate black chalkboard paint, the side of the kitchen unit facing the dining table behaves as an impromptu notebook for the musings and calculations occurring during meal times. Just past the kitchen and dining area, an intimate living room inspires relaxation and conversation. Here, the family members can lounge on one of the two charcoal grey

sofas, play a few melodies on the piano, or take a climb up the ladder to search for their favorite book along the narrow bookshelf unit that runs up the full height of the wall. A cast iron stove heats this social space during the cooler months of the year. With six of the chapel's original arched windows lining the

space, each distinct area of this shared main room enjoys its own natural spotlight.

THE DEMURE AND STREAMLINED EXTENSION MAINTAINS ITS MYSTIQUE BY BUILDING WITHIN THE CHAPEL'S SHADOW. This new wing derives its form and proportions directly from the chapel. Following the angle of the original slate roof, the height of the new structure matches the lines of the existing ridge and eaves. Due to the change in levels across the site, the ample extension still retains a subordinate role to the chapel and neighboring buildings. The new building sits on the existing stone walls that form the divities on the site. These walls are rebuilt as load bearing walls with additional openings to establish links between the higher and lower garden terraces. Working with materials reminiscent of a shadow, the addition's angular silhouette maximizes space while staying as invisible as possible. The key spatial strategy for the renovation and extension resolves the imbalance of the poorly proportioned extant spaces while providing additional bedrooms for the residence. Opening up onto the raised level of the garden, this shadowed area integrates three bedrooms,

The reintroduction of the iconic blackened material within the interior blurs the distinction between inside and out and gives the space a farmhouse quality.

a bathroom, and utility room, each with varying levels of garden access. The utility room, wrapped by a window seat, culminates in a glass corner looking out over the garden. A gracious bathroom with a freestanding tub picks up the same dark wood siding from the exterior extension and uses it as an interior accent. The reintroduction of the iconic blackened material within the interior blurs the distinction between inside and out, and gives the space a farmhouse quality. Simple yet inviting, a rug with warm, rich tones creates a soft transition between the black and white tub and the concrete floor. While bathing, a panel of the black façade can be flipped back to reveal a floor-to-ceiling window to the outdoors. This built-in flexibility allows for more private and more open bathing experiences.

The bedroom wing of blackened timber battens draws from local precedents of black-painted timber barns and the vernacular tabernacle churches. All windows and roof lights align flush with this taut black skin, and an internal gutter allows the building to more closely reference the form of the chapel. The removal of a number of extensions at the rear of the vestry reinstates the chapel and vestry as a conjoined pair of buildings, independent of the stone walls that surround them. In this vein, the new bedroom addition keeps a respectful, detached distance through the implementation of the glazed interstitial passageway. This passage, which can be opened up entirely in warmer weather, connects the extension back to the refurbished mid-19th century spaces, establishing a clear dialogue between past and present. ∎

PLANS & DRAWINGS

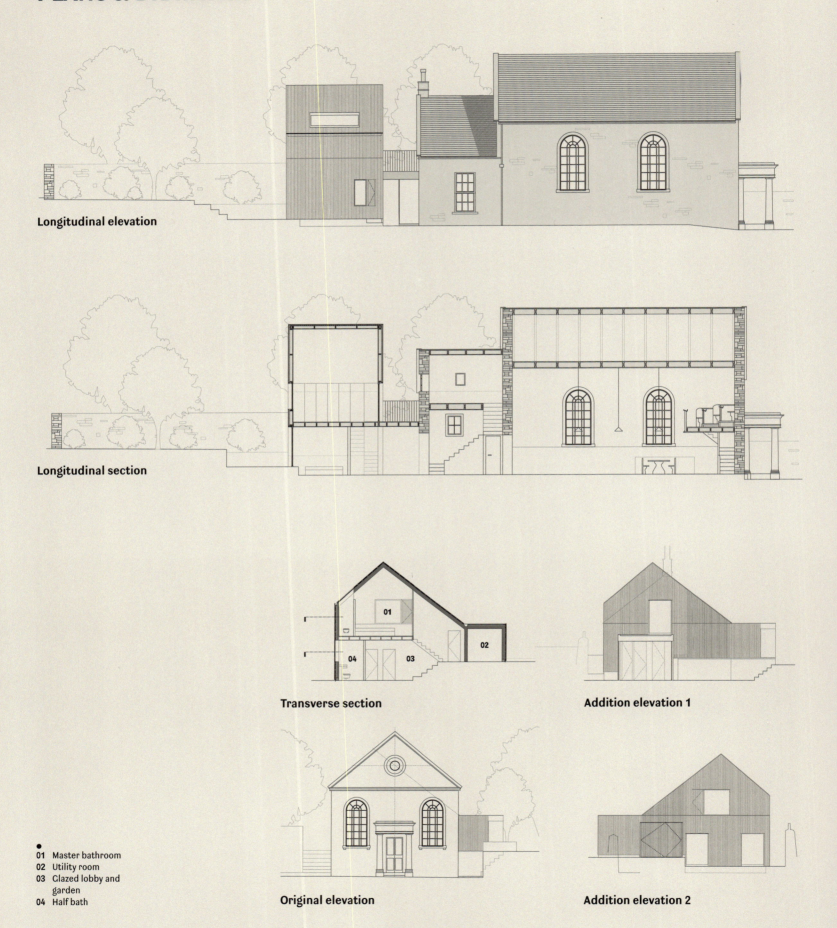

Longitudinal elevation

Longitudinal section

Transverse section

Addition elevation 1

Original elevation

Addition elevation 2

- **01** Master bathroom
- **02** Utility room
- **03** Glazed lobby and garden
- **04** Half bath

SHADOW HOUSE

Existing ground floor plan

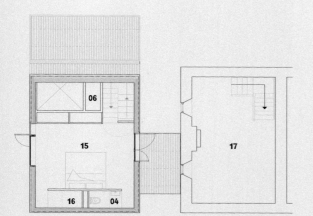

Proposed first floor plan

●			
01	Master bathroom	13	Bathroom
02	Utility room	14	Modern extension
03	Clazed lobby and garden	15	Master bedroom
04	Half bath	16	Shower
05	Carden	17	Guest bedroom
06	Storage	18	Walled garden
07	School room	19	Carden storage
08	Pulpit	20	Bedroom
09	Pews	21	Library
10	Chapel	22	Kitchen garden
11	Porch	23	Kitchen and living room
12	Gallery		

Spatial Strategy

Set in the shadow of an original chapel, a residential extension carefully updates the site for a growing family. The double-height chapel interior converts into an open kitchen, living, and dining space with the bedrooms and bathrooms situated in the addition behind.

●
Number of square meters:
286
Number of rooms:
11
Number of floors:
2
Number of residents:
5
Sustainable features:
Rain water harvesting,
ground force heat pump,
newspaper insulation for internal walls,
eecycled fiber board for internal walls,
locally sourced cedar for external cladding
Outdoor areas/rooftop access:
Backyard
Context:
Village
Type:
Addition and renovation

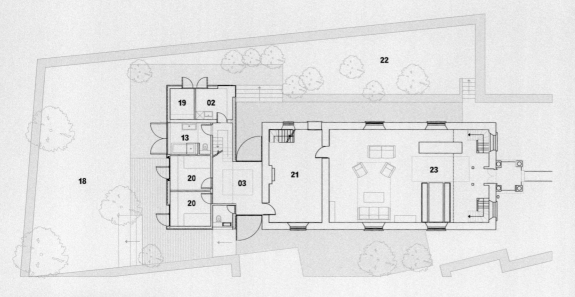

Proposed ground floor plan

Brandt + Simon Architekten
SCHUPPEN
Berlin, Germany, 2009

■ A charming, slender home on a narrow residential plot becomes an eye-catching addition to the neighborhood. Covered in a vibrant skin of geometric shingles, the ten-room house easily supports a couple and their two children. The home, located in Berlin's Pankow district, provides space for the family and their extensive book collection. Bettering both mind and body, the residence includes a large study and an ample outdoor area. This outdoor area comprises a lawn for the children to play as well as a garden where fruit trees and vegetables grow. This valuable access to both green space and homegrown produce nurtures a more sustainable lifestyle within the urban context.

Transitioning from shades of dark green at the bottom to a pale greenish white at the top, the exterior shingles act as a graphic transition between garden and sky. This dappled patterning resembles the intricacies found on the leaves of a tree. **FUNCTIONING AS A TYPE OF UNIQUE CAMOUFLAGE, THE SHINGLE SYSTEM HELPS BLEND THE HOUSE INTO THE SURROUNDING NATURAL SCENERY.** The green home also offers a synthetic feeling of the natural world against the backdrop of the neighboring linear, white houses. Bold yet inherently covert, the verdant graphics fade into the plants and trees that fill the deep gap between the neighboring buildings. This unexpected form of dazzling camouflage effectively conceals the house from the hasty glance of passersby.

The shape of the repeating tiles and their chosen color range develops an exciting interplay between the traditional building materials found in the area and the haptic quality of their pixelated appearance when viewed from afar. This shifting pattern grows and fades in intensity according to one's distance from the façade. Recalling a former nursery on the site, the carefully chosen color gradient

works as a repeating pattern. In addition to behaving as an homage to the past, the exterior approach also responds to the clients' wish to build a true garden house.

This blanket of colorful tiles provides optimal coverage for protecting the timber frame construction just behind. **COMPLETING THE BUILDING'S GREEN THEME, THE PROJECT APPLIES A NUMBER OF SUSTAINABLE ELEMENTS.** First, a building cellulose made from recycled paper serves as the exclusive choice for the structure's thermal insulation. Passive solar energy panels stand as the second sustainable attribute integrated into the home. These solar

lot. Ignoring the classic linear placement, these generous windows strongly relate to the rooms within. Their dark, zinc-plated frames stand out against the green tiles and animate the building with a network of curious eyes. Linking the ground floor living spaces with the outdoors, sliding glass doors expand the interior out onto a wooden garden deck. This new channel between inside and out encourages open air dining during the summer months. Modest playground equipment including a small swing set and a sandbox entertain the children in the garden. The proximity between the terrace and play space allow the adults and children the freedom to focus on their distinct areas while always remaining in visual contact. This terrace also demarcates the more traditional lawn area from the vegetable garden around the back. A series of smaller windows penetrate the side and back façades, behaving as key spotlights for illuminating the interior.

WITH NO ROOM LIKE ANY OTHER, THE INTERIOR BRIMS WITH INDIVIDUALITY. Numerous small rooms work together to form an adaptable internal structure. Many windows extend over the full width of the white interior walls and determine the orientation of each room for maximum exposure to sunlight. Both the kitchen and dining areas look out over a view of the patio and the walled part of the garden. These broad views to the outdoors enhance the feeling of spaciousness inside the modestly scaled quarters. A workroom on the upper level mimics a private treehouse. Surrounded by dense treetops that bleed into the northern sky, this place for focused concentration inspires a meditative and diligent work ethic. Bedrooms and bathrooms are set

panels support the three-story building's heating needs during the coldest months of the year.

The building engages a hybrid roof that mixes a combination of pitched and flat roof surfaces. This more robust choice of roof references the vernacular silhouettes of its row house neighbors. Retaining a largely square footprint that tapers toward the back, the residence enjoys broad exposure to the sun across three of its four faces. Large rectangular windows placed vertically and horizontally capitalize on the ample daylight present on this corner

alight by the morning sun. Ample windows ensure that the children always remain in view as they play in the front yard. The library serves as the only room missing a large window. This introverted decision motivates disciplined and focused work. Instead of windows, the walls are used as bookshelves for the family's cherished library collection. Indirect lighting flows into both this space and the other more private areas of the house, ensuring a high level of intimacy throughout. The atypical home's contemporary, positive, and spirited demeanor enlivens both the family who occupies it and the neighborhood at large. ∎

PLANS & DRAWINGS

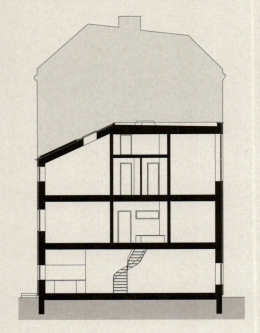

Section 1

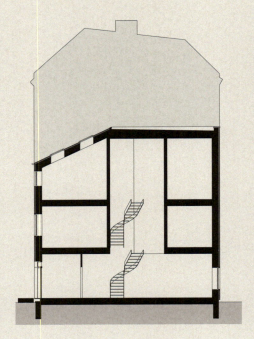

Section 2

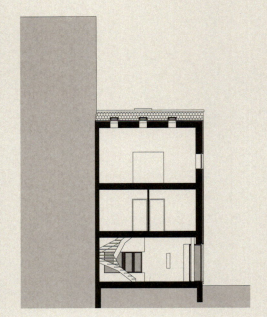

Section 3

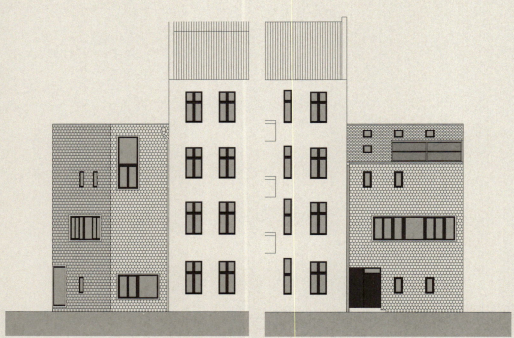

North elevation

South elevation

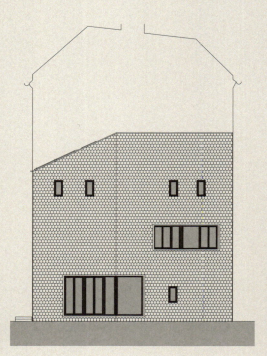

East elevation

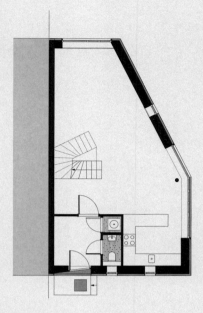

Ground floor plan

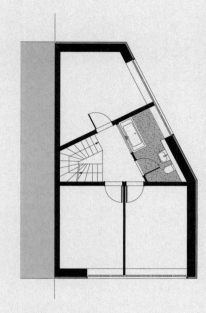

First floor plan

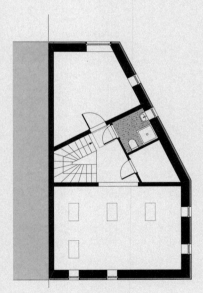

Second floor plan

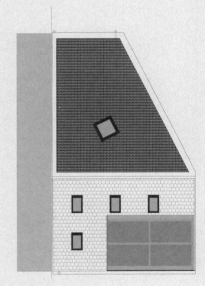

Roof plan

Spatial Strategy

This narrow residence embraces its relationship to the outdoor garden. With public spaces below and private areas above, carefully placed windows ensure all parts of the house receive adequate access to sunlight.

●

Number of square meters:
155
Number of rooms:
9
Number of floors:
3
Number of residents:
2 adults, 2 children
Sustainable features:
Solar hot water,
floor heating system,
gas condensing boiler
Outdoor areas/rooftop access:
Side yard and garden
Context:
City center
Type:
New construction

Braun und Güth
PÜNKTCHEN
Frankfurt, Germany, 2011

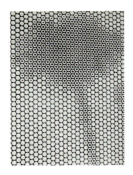

After being relegated to urban anonymity following the aggressive renovations after World War II, a thoughtful and recent renovation brings a turn of the century building back to life. The multi-story residence, formerly dubbed as the ugliest house on the street, now stands as a highly individualized, warm, and bright townhouse in the middle of Frankfurt's urban center. The classical house sits in the northern part of the city in a row with similar neighboring structures. A narrow zone of front yard separates these row houses from the street. Following a total renovation and exacting restoration, the meticulous selection of project partners, procedures, forms, and materials develop an exceptional and customized result.

Despite the decayed state of the building after years of neglect, the clients responded to the traces of its previous grandeur. The drive to revive and enhance the lingering and latent qualities of the home by complementing them with an inviting and contemporary design serves as the key focus of the project. Rather than highlighting contrast, the design approach instead synthesizes a unity from the old to the new.

The classical structure divides into four levels, culminating in an impressive attic studio space at the very top. Massive horizontal Spanish sandstone pieces deepen the façades theme of layering. Other areas of the stone exterior are carved with a fine and delicate ornamental pattern, turning its inherent massive appearance into a nearly textile-like lightness.

Several of these panels on the upper levels showcase computer programmed macro motifs of plants. Oscillating between abstraction and representation, the floral imagery subtly introduces a sense of the natural, organic, and contemporary within the more traditional overarching surface detailing. The striated, grey stone intersperses wood doors and shutters of a golden hue to add a splash of warmth and color to the home's public face. These pockets of sliding shutters afford the interior a flexible attitude towards privacy. Micro balconies with glass guardrails break up the monolithic façade into more accessible intervals while embedding multiple links to the outdoors for the family inside. The diverse outcome inspires a peek-a-boo game between private and public, the hidden and the open. Breaking from the main façade's symmetry as it faces the street, the house opens up to the garden. These different exterior elements defy a strict order, instead working together to form a three-dimensional collage.

Combining the variety of existing and added elements, the interior design applies a reduced and precise selection of materials to achieve a harmonious and comfortable atmosphere. Largely inspired by the minimalist music of José González, the interior utilizes a series of repeating materials that accentuate one another. These material choices include oak, limestone, black anodized steel, and glass. The resulting surfaces radiate a discreet sophistication that help ground the eclectic, and at times opulent, details in a cloak of humility.

Original details remain including old paneled doors and a grand curving wooden staircase. The entry floor features two vaults that offer visual and physical connection to the backyard garden. One of these vaults converts into a spectacular and intimate dining room. A reversal of the original layout ignites a greater involvement in both the communal and outdoor spaces, including the addition of a sun terrace on the attic level.

Two different strategies inform the link between indoor and outdoor spaces. In the first instance, a stone wall screens the narrow front garden from the road. Proudly marked with a single tree, this public outdoor space becomes a closed courtyard. The graceful interior's openness to its garden serves as the second unique spatial approach. Here, the attached two-story winter garden with its delicate olive tree blurs the line between house and garden. Embodying a feeling of utter weightlessness, this glass shell engages a cantilevered structure. The

light and self-supporting structure with a bonded aluminum frame enables the omission of the corner post. This open corner frees up the space for wrap around, double-height glazing for better enjoyment of and access to the lush outdoor garden beyond. **DURING THE COLDEST TIMES OF YEAR, THIS WINTER GARDEN HELPS MITIGATE THE HARSHNESS OF THE CLIMATE, ALLOWING A RARE SENSE OF OUTDOOR ACCESS.**

Rather than highlighting contrast, the design approach instead synthesizes a unity from the old to the new.

The temperate respite maintains a pleasant and inviting temperature throughout the shifts in season. Jutting out from the original building, this north-facing glass house makes space for a generous balcony above.

Occupying the entire south side of the large pitched roof attic space, the studio window on the top floor proves comparatively generous. The individual elements are designed as a sliding glass window, operable via a cabled motor. Built-in furniture, including a discreet bar station, conceals the closed border areas on the space's gabled ends. These wooden storage units blend into the thick plank flooring, resulting in a room where walls and floor read as one. Soft, light grey sofas bend through the space. **WELCOMING IN BOTH SKY AND CITY, THE EXHILARATING ATTIC ENCOURAGES FAMILY GATHERINGS AND CONTEMPLATIVE MOMENTS OF OBSERVATION.** ■

PLANS & DRAWINGS

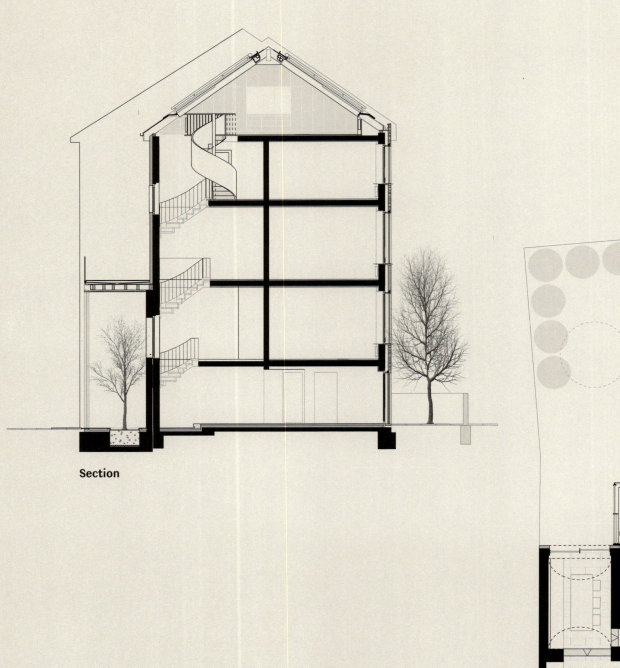

Section

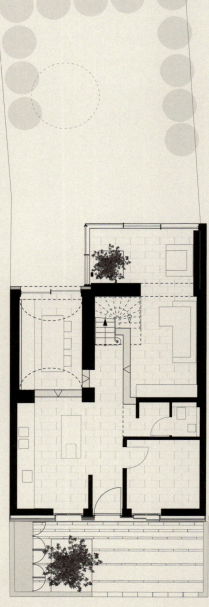

Ground floor plan

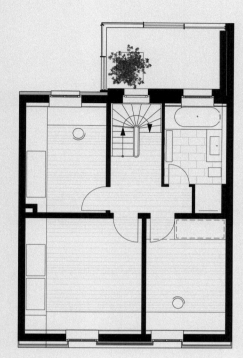

First floor plan

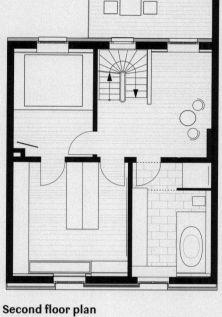

Second floor plan

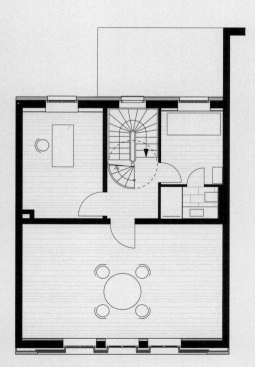

Third floor plan

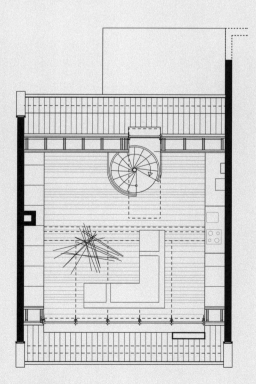

Top floor plan

Spatial Strategy

The five-level townhouse relates not only to its residential context but also to its lush backyard garden. A spiral staircase winds its way up to a transformed attic space at the very top. This spectacularly bright space becomes the family observation room.

●

Number of square meters:
300
Number of rooms:
12
Number of floors:
5
Number of residents:
6
Sustainable features:
N/A
Outdoor areas/rooftop access:
Front and backyard plus roof terrace
Context:
City
Type:
Renovation

C. Fischer Innenarchitekten
TOWNHOUSE
Berlin, Germany, 2010

■ This towering multiple level house caters to the diverse needs of a designer couple in a densely populated area of Berlin. Inspired by the typographer's strict base gridlines, the home takes on a similar linear logic in its architectural approach. The construction faces the standard difficulties of building in the innercity, including a proximity to the subway, with easy strides and appealing solutions. Flanked on either side by

government buildings, the eight-story home works within its dense urban context. Seven of the eight floors of the residence sit above ground while a lower floor below serves as a basement and garage. With personal terraces reaching out from each of the seven upper floors, the front face of this slender townhouse choreographs an engaging relationship with the city. The tall glass structure gracefully relates to its neighbors. Floor-to-ceiling glazing and

sliding glass doors transform the building into a glowing beacon in the evenings.

IN ADDITION TO AIDING THE REPOPULATION OF THE AREA, THE BUILDING ALSO PROMOTES A STREAMLINED APPROACH TO SMALL-SCALE, INDIVIDUAL ARCHITECTURAL EXPRESSION. The systematic and gridded façade steps back on the highest floor to create a penthouse experience with a prominent rooftop terrace. By comparison, the back side of the residence proves exceedingly more private. Here, the building manifests as a more opaque façade comprised of a patchwork of smaller windows that shift in transparency. This rhythmic and abstract fluctuation between transparency and translucency generates a modern yet unique public face. Looking inwards instead of outwards, the back of the house preserves a level of privacy

and discretion for the residents within. Presenting two distinct faces to two distinct streets, the design embeds a sense of directionality into the layout.

Following the basement and underground parking, an entry level leads to the home office that takes up the second and third floors. This office space features a double-height layout with a mezzanine above. Such a special open treatment for the work area encourages an interconnected floor plan that inspires collaborative thinking. The fourth floor houses the home's technical appliances while the fifth, sixth, and seventh make up the main part of the residential living quarters. A small garden completes the upward momentum of the home, introducing a natural element into the manmade structure.

Minimalist, efficient, and simple interiors unite the different levels of the residence. The layout maximizes the narrow spaces via a surprisingly intricate system of shifting partitions, cabinetry, and double-height bookshelves. Raw ceilings,

concrete floors, and plentiful built-in storage in light wood and white panels, orchestrate a humble yet energetic atmosphere. A flexible inner wall lining of individual plates ensures that the underlying structural installations and finishes relate to one another. The stylish but no-nonsense kitchen embraces a stripped down aesthetic sensibility. **UNPRETENTIOUS AND YOUTHFUL, THE INNER WORKINGS OF THIS HOME CELEBRATE THE NUANCES FOUND IN THE EVERY DAY.** With various loft spaces, mezzanines, and interconnected floors, the home offers an engaging diversity and utility from space to space and floor to floor.

Adjustable sliding walls add a level of flexibility to the floor plan. These partitions, embedded into the main walls of

Minimalist, efficient, and simple interiors unite the different levels of the residence.

The home integrates a number of sustainable features. These energy efficient systems work in concert with the modern building envelope to generate a contemporary approach to sustainability. Aside from being ideal for privacy considerations, the translucent back façade applies a system of three-pane insulating glass. This multipurpose glass not only prevents overheating but also shields from glare while providing thermal and acoustic insulation for the interior. Efficient floor heating and a system for solar hot water round off the sustainable offerings. Both client and architect share a fascination with new materials and technologies, and this shared interest appears throughout the final design. The state-of-the-art systems blend technology with functionality to become an integral part of the residential atmosphere.

The residence succeeds in cultivating its own individual character while still echoing the patterns present in the urban environment around it. Like an intriguing chameleon, the townhouse's exterior presence shifts from day to night. Throughout the daylight hours, the residence appears more closed to the outside, protected behind panes of frosted glass. By night, however, the house lights up from the inside—a lantern and wayfinding marker for the neighborhood. ∎

the house, break up or seam together the different spaces as needed. This built in adaptability responds to the changing needs of the clients over time. Fluctuating between open loft spaces or intimate smaller rooms, these walls enable the residence to remain relevant for longer, as the couple can easily expand within its fixed perimeter.

PLANS & DRAWINGS

- 01 Roof terrace
- 02 Bedroom
- 03 Livingroom
- 04 Balcony
- 05 Kitchen
- 06 Utility room
- 07 Guestroom
- 08 Storeroom
- 09 Conference room
- 10 Office
- 11 Garage
- 12 Entrance
- 13 Service connection room
- 14 Bored pile wall

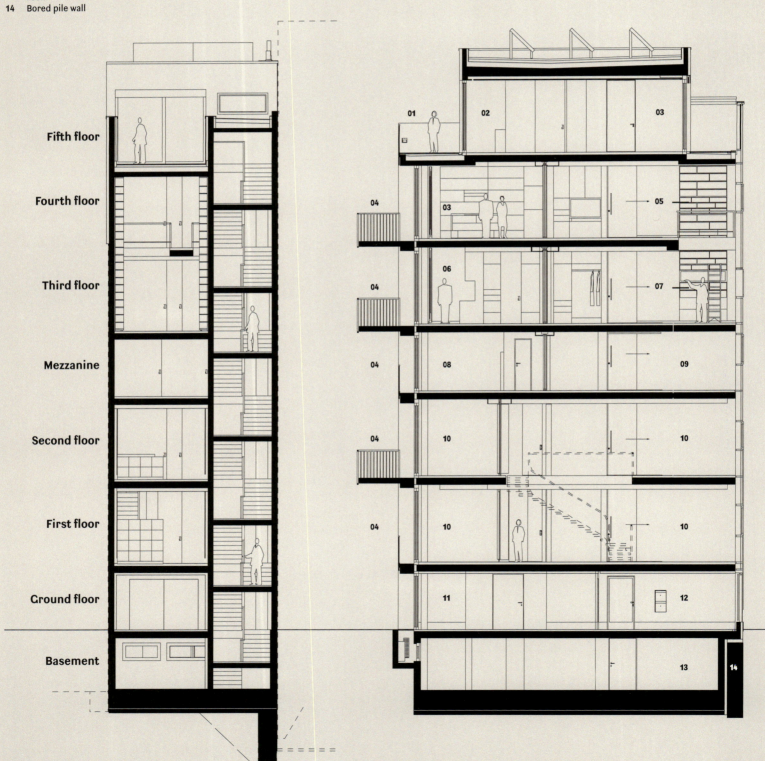

Fifth floor

Fourth floor

Third floor

Mezzanine

Second floor

First floor

Ground floor

Basement

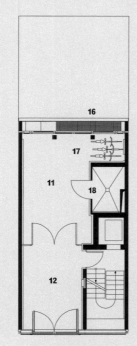

Basement
15
13
14

Ground floor plan
16
17
11
18
12

First floor plan
04
10
19
20
10

Spatial Strategy

An eight floor residence mixes living and work environments. Built-in sliding partitions allow a flexibility between the spaces. With some areas double height and others single levels, the diversity of the layout shifts from floor to floor.

●

Number of square meters:
398
Number of rooms:
6
Number of floors:
8
Number of residents:
2
Sustainable features:
Floor heating system, solar hot water
Outdoor areas/rooftop access:
Garden
Context:
City center
Type:
New construction

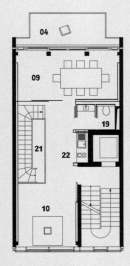

Second floor plan
04
09
19
21
22
10

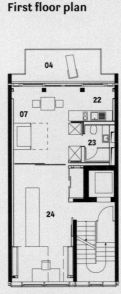

Mezzanine
06
19
08

Third floor plan
04
22
07
23
24

Fourth floor plan
04
05
03

Fifth floor plan
01
02
23
03

●
01 Roof terrace
02 Bedroom
03 Living room
04 Balcony
05 Kitchen
06 Utility room
07 Guest room
08 Storeroom
09 Conference room
10 Office
11 Garage
12 Entrance
13 Service connection room

14 Bored pile wall
15 Server room
16 Yard
17 Bicycle rack
18 Garbage shack
19 Half bath
20 Garderobe
21 Void
22 Kitchenette
23 Bathroom
24 Studio

GAS STATION

Berlin, Germany, 2008

■ Part home, part gallery, this residential project repurposes an old gas station in the Schöneberg neighborhood of Berlin. Dating back to 1956, the gas station remained vacant for decades, sustaining considerable structural damage over the years. The updated and transformed building now serves as the home of a lively Swiss gallery owner and art collector. In spite of all odds, the gas station now stands as the sole freestanding private residence in central Berlin. Seeing the potential of the derelict structure, the courageous client brought in two sets of architects to collaborate on the extensive refurbishments as well as the construction of a new addition.

The addition introduces an impressive gallery building and a crimson red bathroom box. This free-standing cube of a bathroom resembles a giant Christmas present from the outside and continues to dazzle the eye from within. Once inside, guests encounter a vibrant mosaic made from Bisazza tiles surrounding an oversized bathtub that can easily accommodate up to five people. An advanced, energy efficient heating and cooling system runs under the floor of the three spaces.

THE VISION FROM THE OUTSET COMPRISES THE RESTORATION OF THE EXTANT BUILDING AND THE REHABILITATION OF ITS ORIGINAL EXTERIOR APPEARANCE. In addition to adding new interior insulation, the glass tiled façade is now retrofitted with original parts. Old steel doors and the fragile glass enclosure in the former sales room are reconditioned and strengthened, complimenting the small rebuilt steel windows.

With the large part of the interior structure of the existing building preserved, the individual rooms adopt new functions. Here, the former workshop transitions into the kitchen and the sales room morphs into the office and dining room. The tasteful kitchen maintains an old world feeling with glossy tiles, frosted glass, and off-white walls. Professional, stainless steel appliances lead into a more contemporary dining space accented with cherry red window frames

and chair cushions. This bright space opens up to the garden, encouraging an indoor/outdoor dining experience. Floor-to-ceiling bookshelves host the client's extensive library. This diverse collection softens the minimalist language of the interior and lets the personality of the owner shine through. Supported by a selection of artworks on the other wall, the majestic feeling of this library space comes from its direct connection to the outdoors. Full-height glazing

frames the garden beyond. With a crisp white trim on every side and a tree commanding the center of the composition, this window to nature reads as yet another piece in the client's art collection—the ultimate landscape painting.

Positioned directly on the site boundary, the new wedge-shaped, two-story volume extends the existing gas station into a lofty gallery space. Connected together via an internal axis, this new building

offer framed views into the garden and towards the elevated railway nearby. With a tranquil oasis just outside, the new structures imbue an energetic atmosphere to all who encounter them. The pure, clean lines of the addition produce a modern gallery space. Tall, whitewashed walls, semigloss white floors, and abundant natural lighting all work together to craft a neutral backdrop that accentuates the power of the shifting art collections on display.

A high wall encloses both buildings and shelters the new, intimate garden landscape. Designed by Zurich-based Guido Hager Landschaftsarchitektur, the garden boasts fully grown pine and cherry trees, bamboo bushes, and a shallow reflecting pool. Accessed by a steel bridge over water dotted with lily pads, this lush garden seals the site from the hectic

integrates a simple, planar façade. This understated façade creates a canvas-like backdrop to highlight the delicate extant building and its characteristic 1950s gas station roof. The new façade consists of an industrial insulated channel glass system. During the day, this detailing allows for soft indirect lighting to wash into the gallery spaces on the first floor. During the night, this same system enables the building to glow warmly from the inside, revealing its alluring internal structure. Large floor-to-ceiling windows

environment just beyond its perimeter and shields it from direct views from the outside world. The white and yellow flowers and plants elegantly juxtapose against the vintage red tiles of the gas station. Thoughtfully curated, this planting selection keeps something in bloom throughout the year regardless of the dramatically changing seasons. Nine 50-year-old pine trees, the local tree of the area, match the mature nature of the original building. Inspired by the Chinese principle known as "the three friends of winter"—a resilient trio of pine trees, plum trees, and bamboo—the garden exudes a timeless quality. Together, the fresh addition and garden form the framework for the building's second life—a stage for elevating the old gas station into a collector's item and art exhibit in itself. ∎

The vision from the outset comprises the restoration of the extant building and the rehabilitation of its original exterior appearance.

PLANS & DRAWINGS

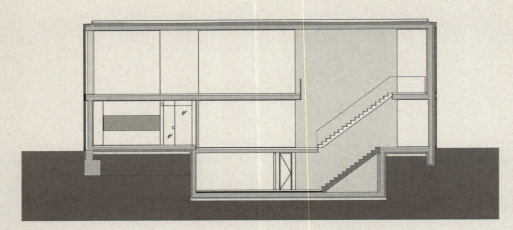

Section

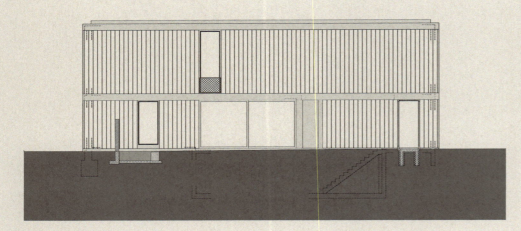

West elevation

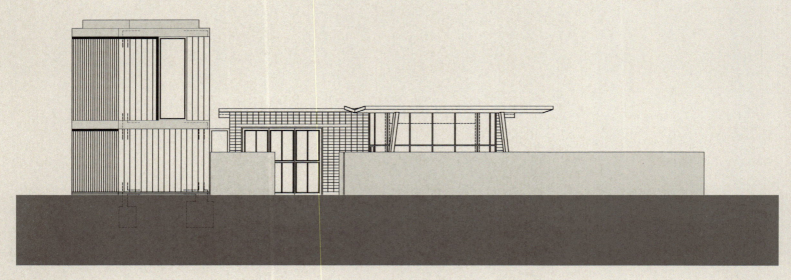

North elevation

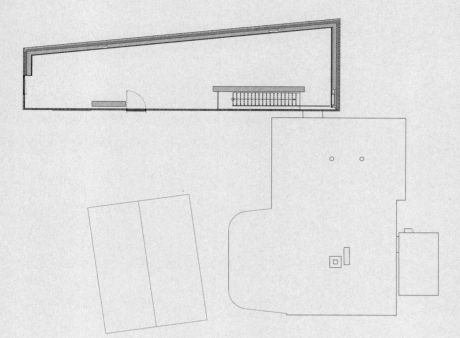

First floor plan

Spatial Strategy

Transforming a 1950s gas station into a residence and art gallery stands as an impressive feat in terms of layout. The original gas station remains intact, housing the residential areas of the project. A double-height cube next door accommodates the client's impressive art collection.

Number of square meters:
340
Number of rooms:
6
Number of floors:
2
Number of residents:
1
Sustainable features:
Efficient insulation, repurposed materials
Outdoor areas/rooftop access:
Garden
Context:
City center
Type:
Addition and renovation

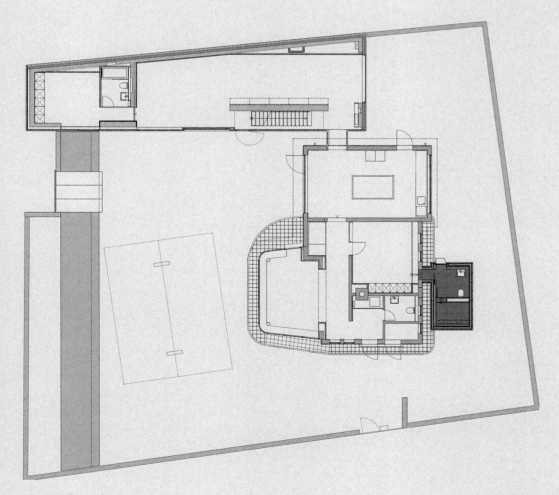

Ground floor plan

Hideaki Takayanagi
LIFE IN SPIRAL
Tokyo, Japan, 2010

■ Developed around the concept of a spiraling porch, this four-story glass home challenges conceptions of public and private living within the urban environment. The porch, referred to as "engawa" in Japanese, plays an essential role in the traditional Japanese house and represents one of its most nostalgic spaces. Located in a dense section of Tokyo, the hefty price tag of the land made it difficult to incorporate the classic porch into the design. Not to be thwarted by the expensive nature of the city, the design approach instead envisions a new version of the engawa. This updated porch tackles the spatial constraints of the narrow site while also providing added benefit for those who live inside. Leveraging the tight plot dimensions, the building's spatial response forms a three-dimensional, spiral-shaped engawa structure.

THIS SWIRLING GEOMETRY FUNCTIONS AS BOTH THE MAIN STAIRS AND THE MAIN ROOM OF THE HOUSE. With the indoor porch set on the outside of the circulation corridor—a white steel ribbon woven through every level—this social area takes on increased functionality and value. The ribbon produces many shaded and hidden spaces both inside and out. These spaces support numerous activities, including a playroom for the children and a sunroom for relaxing. As the width of the engawa proves slightly wider than usual, the residents are able to decorate the charming space with chairs and other furniture.

Painted in a fluorine coating, 4.5 mm-thick iron plates generate all of the exterior finishes, including the four floor slabs, spirals, and roof. These plates, painted instead in a emulsion coating, also become the finish of the interior ceilings. Sandwiched together, the dual iron plates make for a strong and stable structural system. For the glass curtain walls, welded and heat-treated steel, flat bars brace the walls and slabs, crafting a series of extremely smooth and flat surfaces. Traditionally used in Japanese shipbuilding, this efficient technology can be easily transported directly from the factory to the site with the help of flatbed trucks. The ingenious system circumvents the need for builders trained specifically in architecture. Instead, these components can be sourced directly from the shipbuilding factory. **SUCH A FLEXIBILITY IN MATERIAL SOURCING OPENS UP NEW MARKETS AND METHODS FOR EXCITING BUILDINGS ON TIME AND WITHIN BUDGET.**

The light-weight house clocks in at about 40 tons—an average of 30 percent less than its neighbors. Designed to withstand Japan's frequent earthquakes, the monocoque body behaves like an iron ship. A steel wire mesh reinforces the 10 mm thickness of the window glass for further support. With no thick columns needed, all disparate parts unify into a strong and supple configuration.

PRIVACY REMAINS A KEY FACTOR DEFINING THIS GLASS HOUSE. Purposefully offset from the surrounding buildings, each of the home's four floors remain a half step above or below their neighbors. Whether looking in or gazing out, neither sightline ever intersects. This effective choreography keeps the inner workings of the home private. Fully equipped and variable vertical shutters act as further screens to mitigate the family's relationship to the outdoors. These adaptable options allow the residents to place a window or a wall wherever they see fit at the time. **THIS LEVEL OF FLUIDITY AND CUSTOMIZATION ORCHESTRATES A SERENDIPITOUS FRAMEWORK FOR ENGAGING WITH EVERYDAY LIFE.**

The gradual mediation between the private house and the public city rethinks the customs of the urban dweller. In spite of its prominent location and high level of transparency, the residence still succeeds in feeling deeply personal and comfortable. As the brilliant white interior winds its way towards the roof, a series of unique spaces reveal themselves. The interior layout transitions from open living and dining spaces to more intimate bedrooms, a guest room, and

Appearing as a modern ice cube delicately placed between two muscular apartment buildings, the house introduces a moment of serenity and tranquility amidst the bustling heartbeat of Tokyo.

a home office. A striking turquoise blue bathroom dramatically contrasts with the otherwise neutral interior. This bold use of color splashes a sense of whim into the composition.

Appearing as a modern ice cube delicately placed between two muscular apartment buildings, the house introduces a moment of serenity and tranquility amidst the bustling heartbeat of Tokyo. From the street, the house resembles an elegant, glowing gallery space with designer objects and furniture interspersed over multiple levels. These interior pieces exude an iconic quality, standing out against a seamless white backdrop. A concrete-faced ground level

steps back from the main glass façade. The setback and change in opacity enables the building to hang over the site atop this sturdy plinth, amplifying the home's featherweight persona. Accessible from the street, the house has only one parking space. The tiny spot meets the strict Japanese building regulations that limit the amount of width and height allotted to each space. This graceful home overcomes all obstacles to achieve a rare synthesis with the Japanese way of living in and with the city. ∎

PLANS & DRAWINGS

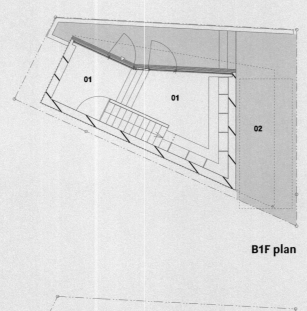

B1F plan

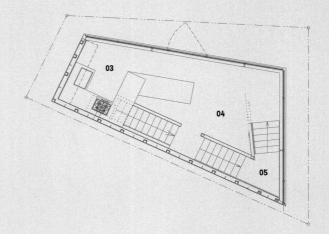

First floor plan

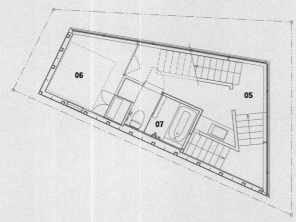

Second floor plan

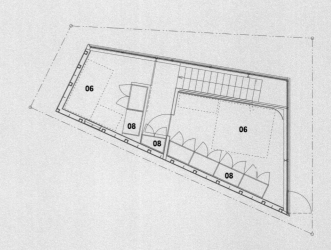

Third floor plan

01 Office
02 Car port
03 Kitchen
04 Living-dining room
05 Spiral porch
06 Bedroom
07 Bathroom
08 Closet

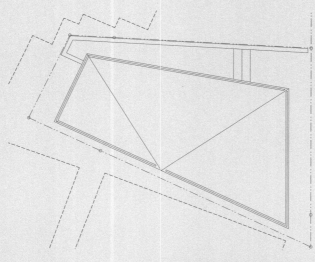

Roof plan (steel origami roof)

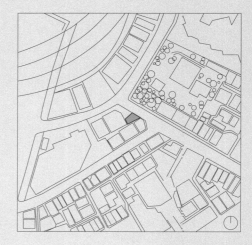

Site plan

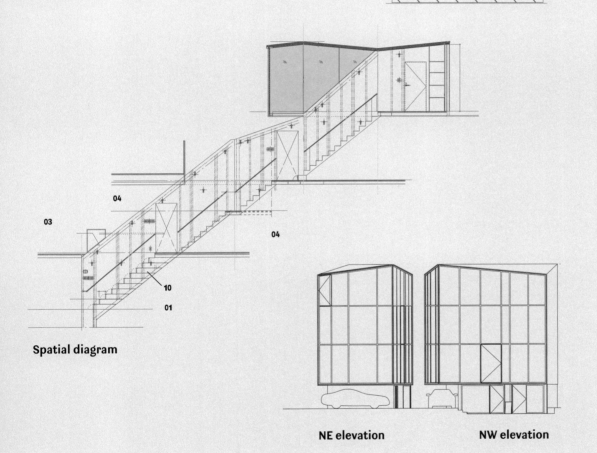

Sections

Spatial diagram

Spatial Strategy

This glass house rethinks the Japanese porch. With a main stairs spiraling across all levels, the interconnected home oscillates between residence and showroom for objects of desire.

●

Number of square meters:
109
Number of rooms:
6
Number of floors:
4
Number of residents:
3
Sustainable features:
N/A
Outdoor areas/rooftop access:
Rooftop terrace
Context:
City center
Type:
New construction

NE elevation   NW elevation

SE elevation   SW elevation

●
01 Office
02 Car port
03 Kitchen
04 Living-dining room
05 Spiral porch
06 Bedroom
07 Bathroom
08 Closet
09 Entrance
10 Shoes shelf

Marc Koehler Architects

HOUSE LIKE VILLAGE

Amsterdam, Netherlands, 2012

Transformed into a residential building some years ago, this vast, elevated home is located in an old harbor cantina on the KNSM-island in the north of Amsterdam. Situated on Levantplein Square, the entirely glazed façades on both sides of the loft feature impressive views out over the water and park. The living patterns of the clients, a young couple awaiting the birth of their first child, directly influence the home's design approach. With both clients participating closely in the project's development, the process began with the couple each describing their version of a perfect day in their ideal house. Building off this data, the architects then mapped the couple's personal belongings and domestic routines. The results of this personal

retain a more intimate and private quality compared to the large, open living room below. The indoor rooftops maximize the functionality of the space without compromising the sense of openness throughout. Benefitting from the floor-to-ceiling glazing on the front and back façades, the intricate yet fluid space fills with sunlight. The method of dividing up the distinct spaces demonstrates the characteristics of a settlement—a group of small houses within the house.

MIXED-USE OPEN SPACES, PLAYFULLY REFERRED TO AS STREETS, WIND THROUGH THE DENSER PROGRAMMATIC AREAS AND TEMPORARILY ACCOMMODATE AND SUPPORT THE VARIOUS RESIDENTIAL ACTIVITIES TAKING PLACE OVER BOTH LEVELS.

input include a spatial frame of thickened walls that function as both a load bearing structure and storage space. Whether dressing up or ironing clothes, closets and cupboards hidden in the walls support the activities and rituals of daily life.

An upper level interweaves another layer of activities into the program. Here, activities ranging from dining, cooking, and office work are located on mezzanines known as roof terraces. These lofted spaces

These small streets emerge as multifunctional living spaces for the more spontaneous, footloose actions of playing, partying, washing, and working. In addition to providing striking city views, the streets also carry daylight right into the heart of the house. Introducing a level of contrast, the house-like volumes contain far fewer mobile activities such as the bedrooms, bathroom, and storage. The compression of certain spaces, including the bedroom, allow others to generously expand. These open spaces can

be colonized in the future by constructing extra volumes when the family grows larger.

Exposed steel beams painted white lend the interior an industrial warehouse aesthetic. A scattering of rectangular skylights brighten the upper mezzanines and contrast with the darker precast paneling of the ceiling. Light wood finishes with prominent natural patterning appear on the sides of several of the raised islands. These warmer surface treatments subtly obscure doors, closets, and other built-in features. Continuing the trend of multi-functionality, the main wooden stairs doubles as a bookshelf underneath. By uniting two core elements into a single gesture, the living room gains additional space. Airy and minimal, the double-height living room centers around a large, L-shaped couch. From this inviting vantage point, the family can observe the happenings of the city through the impressive panoramic windows. A modern take on the cast iron stove floats above the ground and creates a cozy ambiance during the cold

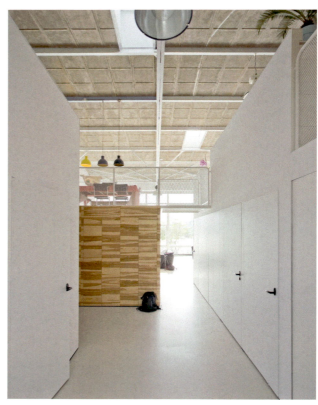

months of the year. Industrial style lights hang down from the rafters across the long floor plan to illuminate the various living and work areas.

The high loft space integrates floor heating to maintain a comfortable climate when it gets cold outside. Ventilators on the ceiling ensure efficient air circulation and prevent it from creeping into the ridge of

and vertical vistas support a cross pollination between programs and family members. Independent but interconnected, the different terraces and lower street levels still find myriad ways to relate to one another. From catwalks to staircases, the relationship from top to bottom, and from one side to another remains in a state of constant dialogue in spite of the complex distribution. A place for

the roof. The closed volumes are insulated and separately heated. Within these more private volumes, the climate can be manually adjusted, creating a dual climate zone.

When both the upper and lower levels of the house are utilized simultaneously, the interior experience recreates the sensation of inhabiting a micro city. With activities occurring across multiple public and private spaces on varying levels, a dynamic rhythm for daily life develops. The flexibility and intricacy of the layout not only allows for adaptation and expansion to meet the family's changing needs over time but also encourages interaction and exchange between different spaces. Broad horizontal

From this inviting vantage point, the family can observe the happenings of the city through the impressive panoramic windows.

private introspection, public socializing, and acts of togetherness, this expansive residence behaves as an intimate village in sync with the nuances and desires of its occupants. ∎

PLANS & DRAWINGS

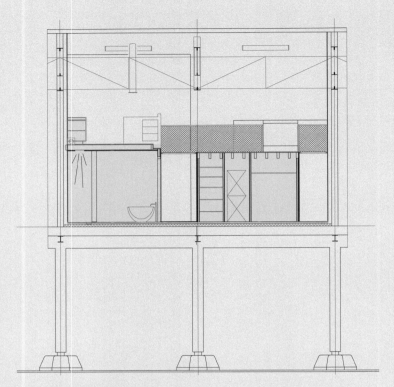

Transverse section

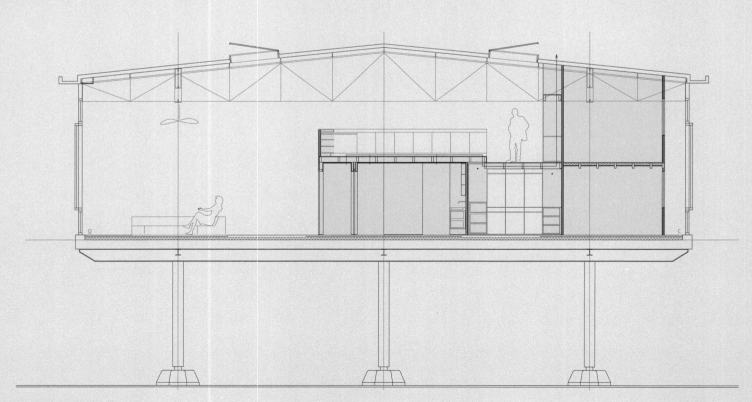

Longitudinal section

HOUSE LIKE VILLAGE

Balcony diagram

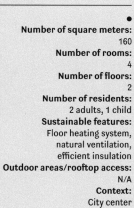

Spatial Strategy

Raised one level off the street, this expansive loft space consists of an array of mixed-use open spaces intersected by denser programmatic areas. The denser blocks of program house the more private areas. The closed boxes make room for an interconnected mezzanine level above.

●

Number of square meters:
160
Number of rooms:
4
Number of floors:
2
Number of residents:
2 adults, 1 child
Sustainable features:
Floor heating system, natural ventilation, efficient insulation
Outdoor areas/rooftop access:
N/A
Context:
City center
Type:
Renovation

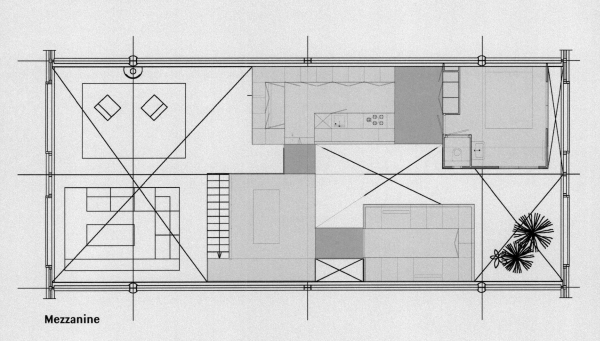

Ground floor plan

Mezzanine

Alphaville
NEW KYOTO TOWN HOUSE
Kyoto, Japan, 2010

■ This three-story home resides on a typical long and narrow site in the centre of Kyoto, the old capital of Japan. Rows of traditional wooden townhouses define this area. Inheriting the advantages of conventional townhouses, the project overcomes their drawbacks through the creation of a more comfortable and

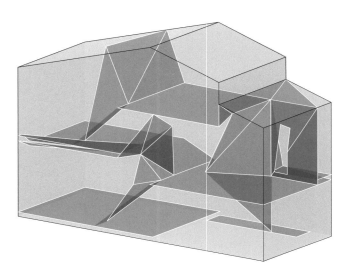

enjoyable space. The polyhedral form of the interior partition walls embody the home's most standout feature. Seemingly designed through intuition, these sculptural forms in fact derive their shapes and gestures from logical concepts in order to perform multiple functions.

The first asset of these unusual partition walls stems from their multi-dimensionality. Rather than conventionally extending in the vertical and horizontal directions, these 3D walls loosely connect the distinct spaces over three floors. The bold spatial result behaves as a continuous room with dynamic nuances.

These articulated surfaces produce an atmosphere that is simultaneously spacious and heterogeneous.

A second feature of these interstitial walls comes from their ability to act as reflectors of daylight. As most typical townhouses cannot afford to have large openings on both the short and long sides of the building, the majority of these projects end up dark and with poor communication between levels. In this project, however, its steel rigid frame and the insertion of light colored polyhedral partition walls overcome the isolated, cave-like quality of this residential typology. Generous openings in the walls and floors, along with the geometric connecting walls, allow natural light to diffuse multi-directionally while encouraging vertical and horizontal communication and movement. The design, freed from the constraints of the old system, inspires the occupants to engage in various relations with each other across the 12 spaces and various levels. Through this way of behaving and relating to one another, a new lifestyle emerges within this historical area. The angled surfaces, made from a creamy Japanese linden plywood, softly reflect the sunlight coming in from both the north and south sides of the building. **BRIGHTENING AN OTHERWISE DARK INTERIOR, THESE CONNECTIVE ELEMENTS PROVIDE A LUMINOUS QUALITY THAT REACHES THE VERY CORE OF THE HOME.**

Finally, the partition walls blur the boundary between architecture and furniture. This grey area stimulates the perception and behavior of these transitional pieces. Melting into floors and ceilings, the plywood-finished walls offer enjoyable and tactile interactions as people pass by, through, and over top of them. Similar to playground equipment, the utility embedded within these dynamic shapes manifests as a contemporary interpretation of architect Le Corbusier's concept of a house as a machine for living.

As elegant as the house appears, it faces numerous challenges with the site. Due to the landscape regulations and the physical context of the neighborhood, the project inherits the traditional form and composition of the classic townhouse. The design of the home diligently meets the requirements of the site's exterior restrictions and fire protection guidelines. This preservation of the traditional Japanese wooden exterior stands as one of the key issues at odds with the area's fire regulations. The integration of a parking space for an electric car inside the house represents another challenge resolved with an appealing solution. Utilizing plenty of high-performance insulation, the house also showcases ample, southern-facing openings to soak up the afternoon sun. This south side of the home accommodates a mandated 40-percent void which is shared by all the residents on the block.

Much of this house's power stems from its dramatic use of contrast. Unlike the bright, light-filled interior—accentuated by maple flooring and white plasterboard—the exterior presents a stealth black treatment of its classically inspired façade. Lightweight roofing tiles sit atop dark galvanized steel with an aluminum sash. The ground level introduces floor-to-ceiling opaque panels that transmit

The partition walls blur the boundary between architecture and furniture.

light into the interior while still retaining privacy from the street. As the building rises, three horizontal strips of windows direct light into the upper floors. The two lower rows of windows are shielded by regularly spaced battens that function as both privacy screens and buffers against solar gain. Set back from the rest of the building, the top row of windows provide a lookout point for observing the happenings on the street from this top floor hobby room. **DURING THE EVENINGS, THE PROMINENT EXTERIOR TRANSFORMS INTO A WARMLY GLOWING LANTERN. THREE OVERHANGS ON EACH LEVEL PICK UP GOLDEN LIGHT FROM THE WINDOWS BELOW, SETTING THE BUILDING AFLAME.** Built for a couple and their child, the design of this residence defies expectations while keeping within conventions. Visionary yet rooted in tradition, the home develops a hybrid language for responding to the architectural lessons of the past without sacrificing a contemporary approach toward the future. ∎

PLANS & DRAWINGS

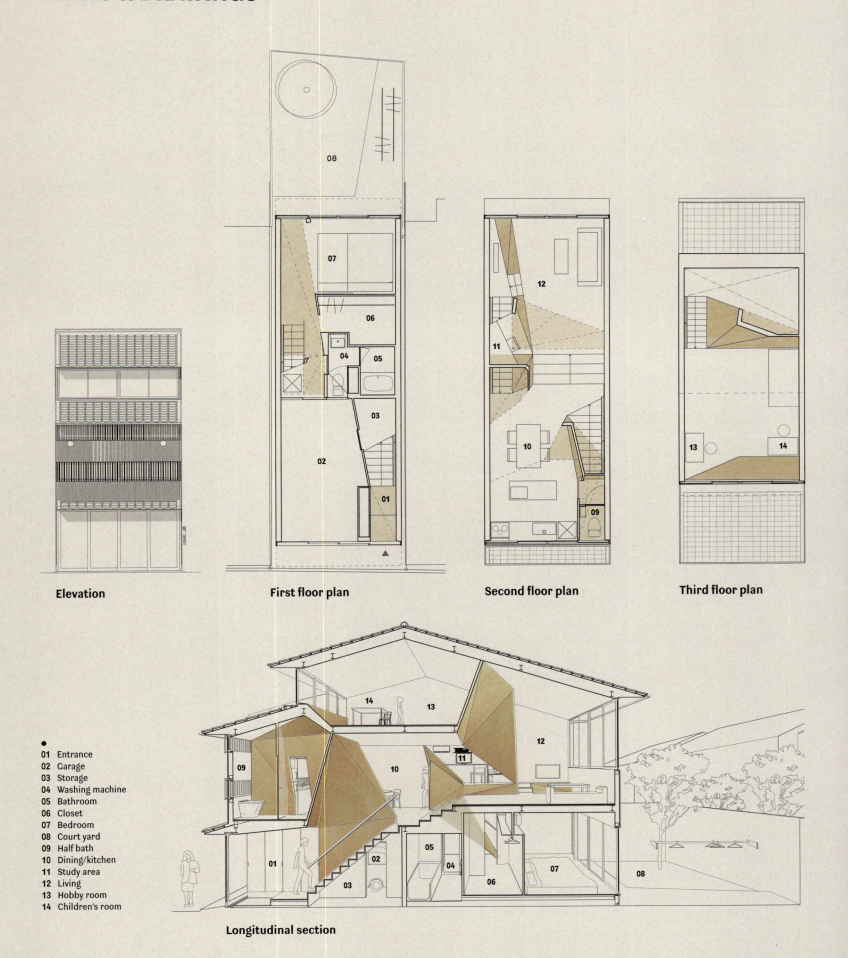

Elevation

First floor plan

Second floor plan

Third floor plan

●
01 Entrance
02 Garage
03 Storage
04 Washing machine
05 Bathroom
06 Closet
07 Bedroom
08 Court yard
09 Half bath
10 Dining/kitchen
11 Study area
12 Living
13 Hobby room
14 Children's room

Longitudinal section

NEW KYOTO TOWN HOUSE

Site plan

Structural diagram

Spatial Strategy

Angular multipurpose wooden panels act as a connective tissue between the three levels of this house. Doubling as furniture and reflectors of light, these interstitial spaces keep the house in a state of constant motion.

•

Number of square meters:
104.66
Number of rooms:
6 (12 spaces)
Number of floors:
3
Number of residents:
2 adults, 1 child
Sustainable features:
Efficient insulation,
southern exposure
Outdoor areas/rooftop access:
An outdoor space is shared by
all the residences on the block
Context:
City center
Type:
New construction

Spatial diagram

Interstitial panel diagram

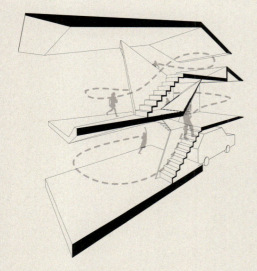

Circulation diagram

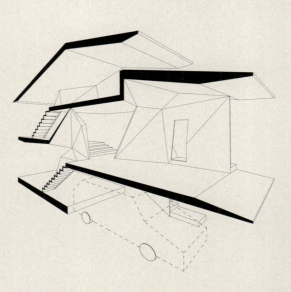

Use diagram

J. MAYER H.
OLS HOUSE
Near Stuttgart, Germany, 2011

The four-person family residence divides itself into an elevated ground floor with entrance area, utility room, and spa, and a second floor with an open, flowing floor plan containing the living, dining, and kitchen areas. Full-height glazing provides unobstructed views of the valley and a generous terrace overlooks the large garden area in the back. The wrap-around windows lining the kitchen are shielded by a fine translucent white curtain. This gracefully thin veil lets in an ethereal, even light and abstracted views of the outdoors while maintaining utmost privacy for the family. Minimally detailed, the kitchen features a streamlined workstation island with appliances and storage sleekly embedded into the back wall. A dark slate counter top and black leather dining furniture juxtapose with their white monotone surroundings. Continuing the futuristic theme with an underlying retro quality, a recessed ceiling with rounded corners appears over the kitchen island and dining area. This ample

■ This unusual hillside home embodies Jurgen Mayer's futuristic, black and white aesthetic agenda. The house rests on a plot of land on a hilly tract near Stuttgart with a generous view of the valley. At the owners' request, the home incorporates these panoramic views into its spatial strategy. A generous backyard and sweeping vistas compensate for the nearness of the home's neighbors on either side. From certain exterior angles, the graphic accents and general composition of the windows and glazing resemble a smiling character. Whimsically shaped picture windows with spotted grey background highlights

speak to the classic camper design language of the past while a glass façade on the second level gives the exterior façade an inviting quality. These unexpected details enliven the house and encourage it to adopt its own distinct character. The house is built in a residential area brimming with conventional developments dating back to the 1960s. Embracing the spaceship style of its design, the boxy yet sinuous home cannot help but stand out.

void adds depth to the ceiling and heightens the spatial experience of the kitchen. Built-in lighting and directional spots introduce a diffused glow over the space. Rhythmic, slender paneling links the kitchen to the living room on the other side of the stairs. This airy living space shares similar recessed ceiling details as the kitchen and orients a family of demure leather sofas and love seats outward to face the garden.

OLS House   195

Embracing the spaceship style of its design, the boxy yet sinuous home cannot help but stand out.

The top level holds the dressing rooms, bathrooms, and sleeping areas which open up onto the outdoor terrace with built-in seating. Situated below the sloping roofline, the surreal bathrooms capitalize on their top floor location with two generous skylights that bring in plentiful sunshine. A sculptural spiral staircase serves as the centrally placed design element connecting all three levels and unifying the matte white interior. All furniture keeps with a neutral color palette. This consistency in tone promotes an interior that remains as fluid as possible. **THE TEMPERED PALETTE ENSURES THAT THE LUSH OUTDOOR SCENERY AND CITY VIEWS ARE NEVER ECLIPSED BY ANYTHING THAT MIGHT DISTRACT THE EYE IN THE INTERIOR.**

A low barrier wall, originating in the backyard, winds its way protectively out onto the street to begin the home's dramatic entry sequence. This curving exterior wall widens towards the street to hold an assortment of plantings and trees. Upon reaching the driveway, the wall wraps inward, leading visitors towards the house. The wall culminates near the front door and raises up a level to proudly showcase a giant "11"—the home's street number—cleanly engraved into its surface. Noticeably different from its light colored upper floors, the lower entry level utilizes an

entirely black color scheme. This band of darkness aids in the illusion of the house appearing to float off its foundations. By creating a relatively inconspicuous entry point with few windows, this street level part of the home can enjoy an element of privacy from car traffic and passersby.

Built as a reinforced concrete construction, the exterior of the residence consists of one heat-insulating compound system, known as EIFS, and an aluminum and glass façade. The choice of using EIFS results in an aesthetic and multipurpose exterior solution, as this cladding system doubles as both insulation and a final finish. Slats and anti-glare sheeting ensure integrated sun protection while also blocking excessive solar gain during warmer months. All of the lightweight partition walls inside are made of drywall, while the floor consists of a seamless layer of screed. The thinness of the roof makes it look more like skin on an animal than a roof on a house. This subtly rounded pitched roof with the deep, recessed balcony engages pre-weathered, zinc plate cladding and is outfitted with solar panels. Combining hard angles with soft filleted edges and geometric cutouts, the home takes on a commanding attitude as it catapults this quiet neighborhood into the future. ∎

PLANS & DRAWINGS

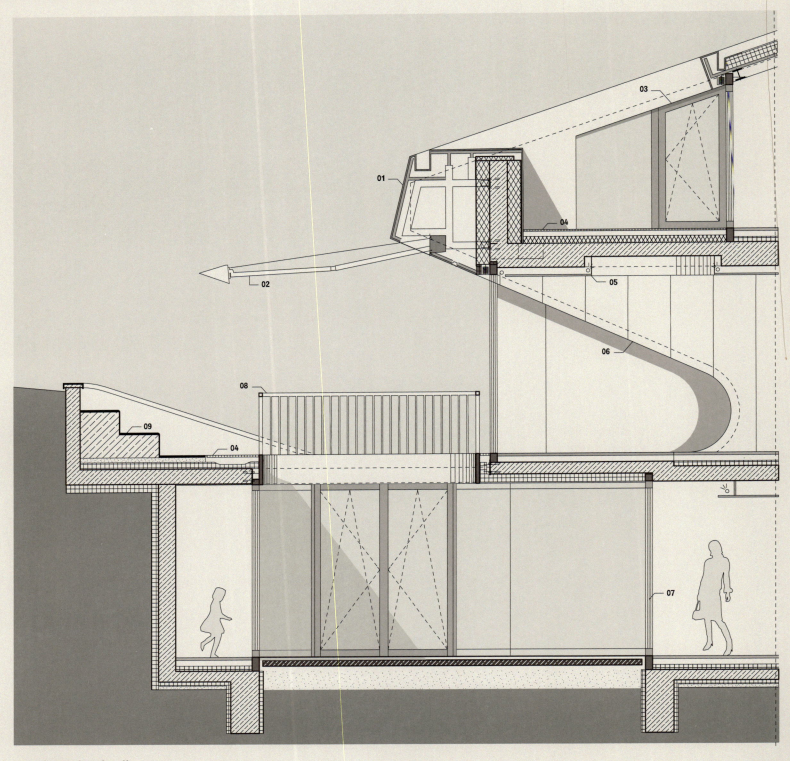

Façade section detail

● 01 Roofing cladding with weathered zinc panels
02 Overhanging canvas blind
03 Rooflight
04 Colored fiberglass grating
05 Recessed lighting
06 Fixed glazing with lamella curtain

07 Inner court façade
08 Courtyard balustrade, colored
09 Terrace flooring, synthetic granular material

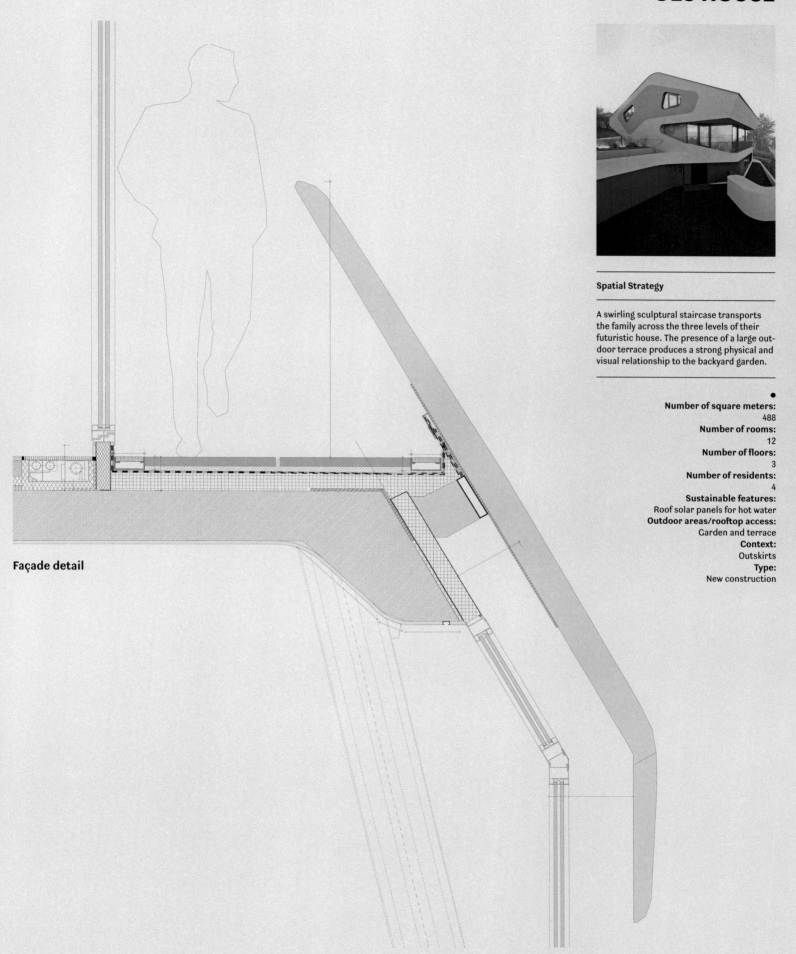

Spatial Strategy

A swirling sculptural staircase transports the family across the three levels of their futuristic house. The presence of a large out-door terrace produces a strong physical and visual relationship to the backyard garden.

●

Number of square meters:
488
Number of rooms:
12
Number of floors:
3
Number of residents:
4
Sustainable features:
Roof solar panels for hot water
Outdoor areas/rooftop access:
Garden and terrace
Context:
Outskirts
Type:
New construction

Façade detail

LIGHT WALLS HOUSE

Surrounded by the charm of old world Japan, a pristine white, double-height box rises into the dense urban fabric. The crisp modern residence enjoys an historical setting brimming with story and tradition. Built for a couple and their two children, a boy and a girl, the house straddles a convenient metropolitan area with busy thoroughfares on one side and the nostalgia found from the family shopkeeper and craftsman on the other. Small wooden houses with classic Japanese tiled roofs line the narrow alleys.

True to its name, light represents the driving concern facing the design approach for the house. The residence succeeds in improving an otherwise completely shady site. In addition to sitting in the shadows of a closely placed two-story house to the south, the winter months leave 80 percent of the plot in cold, dark shadows. At the same time, the very public lot requires a level of privacy for the family. This dilemma posed by the clients to bring the sunlight in while keeping the eyes of the city out stands at the heart of the building's final formal manifestation.

The resulting design develops a luminous space uniformly distributed with sunlight. By controlling the access and direction of the light, a diversity and richness seeps into all aspects of the home. This powerful connection to such a precious natural commodity positively influences the residential spaces, the interior activities, and the family's relationship to the surrounding environment. Striated apertures break up the wooden mass of the roof. **THESE NUMEROUS SKYLIGHTS CHANNEL THE NATURAL POTENTIAL OF THE LIGHT AND GRACEFULLY GUIDE IT THROUGH THE INTERIOR.** Running along the edges of the 9.1-square-meter roof, the skylights form a brilliant outline between wall and ceiling.

An intricate network of roof beams concentrates the sunlight into slender strips. From here, the slightly angled clapboard interior walls—finished with laminated wood—reflect and diffuse the light. As a result, soft and uniformly distributed daylight warmly illuminates the entire spaces.

Workspaces including the kitchen, wash area, and study are oriented underneath the edges of the skylights. Lining three sides of the primary box these more open but distinctive areas bleed into one another as one traverses the perimeter. Embracing the horizontality of the walls and the light seeping in from above, the simple bespoke furniture enhances the sense of directionality across the interior. Made from basswood plywood, a similar

material logic informs the variations in design. These clean and understated surfaces begin as desks and bookshelves in the study area only to later morph into cutting surfaces and storage in the kitchen.

LIGHT, RHYTHM, AND SCALE WORK TOGETHER TO EXPAND THE RANGE OF POSSIBILITIES FOR OCCUPYING AND ENGAGING WITH THE MODESTLY SIZED LIVING SPACES. Defying classic layout conceptions, the house behaves more as a communal interior courtyard than as an assortment of rooms with clearly defined functions and privacy parameters. The novel set up of arranging smaller boxes inside one large box cultivates new uses and activities for both.

The more private bedroom and storage units move deeper into the space and nestle within four white boxes. Each box balances within the larger spatial volume. Wooden ladders above the two storage cubes lead up to two individual bedrooms. These lofted and ultra private rooms encourage new ways of interacting with and perceiving the different levels and spaces of the house. In between these boxes, the remaining areas function as paths and public gathering places. With each box treated as an autonomous structure, the leftover interstitial plaza recalls a bustling small town bathed in light. These empty spaces produce a shortening or elongation of distances between people as they move through the house. When the side corridor opens up, the intermediate spaces connect to the outdoors, generating a thriving social structure between the family members. Potted green plants and hanging vines add a touch of vibrant green into the interior.

From the exterior, the residential box consists of a series of evenly sized white panels. A slight overhang marks the home's main entry point. This overhang lines the track for a sliding, all-white front door. When opened, a glimpse into the warm, wooden interior appears. This corridor space connects to the outdoors on both the front and back façades. The built-in flexibility of this system enables the covered entry space to double as a driveway and parking space for the family's motorcycle. A bed of large rocks that support plant life frame the boundary of the residence. By introducing a rugged and natural element into the composition, the house takes on a lighter and more futuristic character.

The beauty of this residence derives from its simplicity. Merging the natural and universal component of light with the social component of family, the home responds to the nuances of both with great care and foresight. As the sun tracks through, new values and rituals gradually come to light. ■

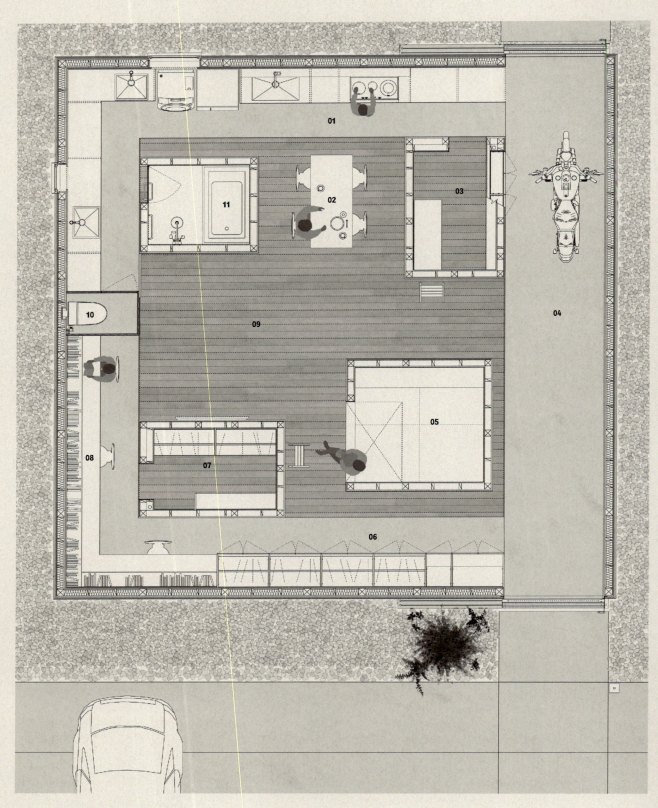

First floor plan

- ● Kitchen
- 01 Kitchen
- 02 Dining
- 03 Storage 1
- 04 Earthen floor
- 05 Bedroom
- 06 Entrance
- 07 Storage 2
- 08 Study
- 09 Living
- 10 Half bath
- 11 Bathroom

LIGHT WALLS HOUSE

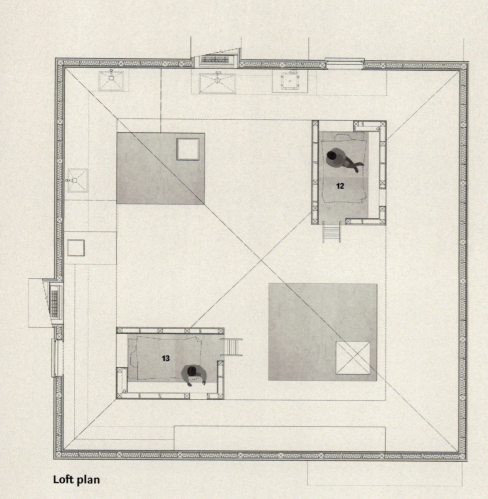

Loft plan

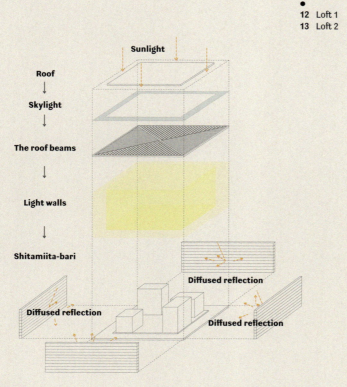

Sunlight diagram

Sunlight

Roof
↓
Skylight
↓
The roof beams
↓
Light walls
↓
Shitamiita-bari

Diffused reflection

Diffused reflection

Diffused reflection

●
12 Loft 1
13 Loft 2

Spatial Strategy

Rejecting the idea of a conventional layout, this space comprises a series of boxes set inside of one large box. With public programs wrapped around each interior wall to enjoy the illumination of the skylights above, the storage and bedrooms remain safely hidden within the solid cubes dotting the center of the space.

●
Number of square meters:
82.81
Number of rooms:
5
Number of floors:
2
Number of residents:
2 adults, 2 children
Sustainable features:
N/A
Outdoor areas/rooftop access:
Rooftop
Context:
Residential suburbs
Type:
New construction

OUTSIDE IN

Takeshi Hosaka

Yamanashi, Japan, 2011

A commission to enlarge the existing house for a young couple in their 30s and their three daughters leads to this charmingly iconic residential design. Built in the Yamanashi Prefecture, the region comprises pockets of one and two-story houses standing close together with empty lots, farming fields, wooded areas, and unpaved roads sporadically interspersed between. Part town and part country, this unusual hybrid setting's bounty and generosity inspires the design of this home. The formal approach of the house choreographs a gradation of scenery from outdoors to indoors.

THE UPDATED HOUSE SUPPORTS THE FAMILY'S INTEREST IN LIVING IN HARMONY WITH NATURE AND THE CHANGING CLIMATE. Even within this crowded residential area, the sound of birds can be heard calling out in the early mornings and sightings of wild pheasants, peafowl, and raccoons introduce a captivating, wild quality to urban life. Featuring an eye-catching sawtooth-shaped roofline, the single-story residence provides a continuous link between the wooded area located on the south side of the house and the interior spaces. The home acts as an open and welcoming boundary structure that subtly marks the separation between the natural and manmade worlds.

With sky above, ground below, and forest and mountains on either side, these natural elements frame and extend into the home's interior. The shell of the house vertically and horizontally incorporates nature as an integral part of the design and develops a gradual transition between inside and outside areas. Completely opaque on three sides, the house opens up entirely along its southern face. This enhanced directionality of the exterior makes the house read as a single room united with the wooded outdoor area. The top part of the residence also engages an open structure that consists of a combination of reinforced concrete V-beams and transparent acrylic. As a result of this unconventional and sculptural solution, the residents can see the sky above. The V-beam structure conveys an impression of durability and reliability, while the transparent acrylic skylights all but disappear into the open air. By only

The formal approach of the house choreographs a gradation of scenery from outdoors to indoors.

perceiving the weight of the beams, the regularly spaced strips of acrylic appear as gaps between the concrete. These prominent breaks in the interiority of the house bring the sky in and up close.

THE HOME TRANSITIONS FROM A SOLID WHITE EXTERIOR TO A TACTILE MIX OF CONCRETE AND BRIGHT WOODS ON THE INTERIOR. A vibrant wood with a yellow tone frames the two layers of windows and doors into the house. These rich wood highlights add contrast against the monolithic nature of the concrete enclosure. The same light wood gradually permeates the interior, extending into the kitchen island and culminating in complete walls where the more private rooms appear. Wooden flooring carefully demarcates the warm children's area from the continuous kitchen and living spaces. This unconventional trio of built-in micro bedrooms consist of a shared play area that bleeds into the main part of the house. Modest curtains appear on each of the two tiers of bunk beds. Evoking the language of a train's sleeping compartment, these simple curtains serve as the sole dividers between the different children's private spaces. This openness promotes a sense of community and connection between the siblings and their parents.

DURING THE LONG WINTER MONTHS, THE HOUSE RETAINS A PLEASANT AND ACCESSIBLE RELATIONSHIP WITH THE OUTDOORS. The open design encourages the family to enjoy the generous views and connection to nature without needing to leave the comforts of their cozy home. Cries of wild pheasants still echo in the morning air, while peacocks can be spotted from the bedroom. In the dining area, the plants and trees continue to flourish and act as a mini oasis during the coldest days of the year. Throughout the hot summer months, the house relies on natural ventilation instead of air conditioning. The two layers of windows at the front of the residence can open to their full width to circulate fresh air and synchronize the interior and exterior temperatures. Even on rainy days, the family can leave the windows of the dining room open to irrigate its flower bed floor.

The dining room, located at the end of the house, behaves as the boundary area between the outside and inside. Still technically inside the house, this indoor/outdoor dining space opens up completely to the outside. Set on the ground among plants,

flowers, and trees, and placed directly under the final rectangular skylight of the house, the dining area feels more like a secret garden than an indoor space. An organically shaped stone island hosts the intimate dining table and chairs. Complete with a hammock for gazing at the stars, the dining space is fitted with floor-to-ceiling folding glass panels. These panels transform this multipurpose space into an outdoor terrace or a more private indoor greenhouse, depending on the mood of the family and the prevailing climate. Cut off at the highest point of the zig-zag roof silhouette, the truncated volume reaches out in a bold and friendly gesture to the surrounding nature. ■

PLANS & DRAWINGS

Section

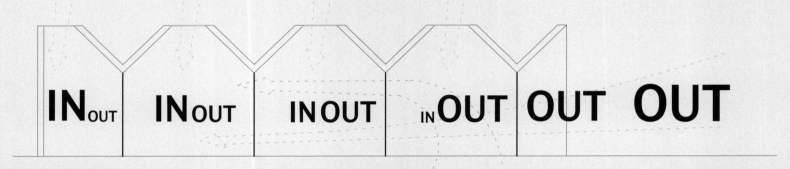

Concept sketch

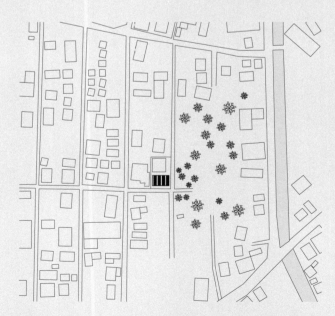

●
01 Closet
02 Children's room
03 Living
04 Dining

Site plan

Spatial Strategy

True to its name, this house moves from the outside to the inside. Closed on three of its four sides, the layout frames an outdoor garden. The dining area serves as the transitional space where home and garden blur.

Number of square meters:
102.14
Number of rooms:
5
Number of floors:
1
Number of residents:
2 adults, 3 children
Sustainable features:
Efficiently insulated acrylic
Natural ventilation
Outdoor areas/rooftop access:
Garden
Context:
Outskirts
Type:
Addition and renovation

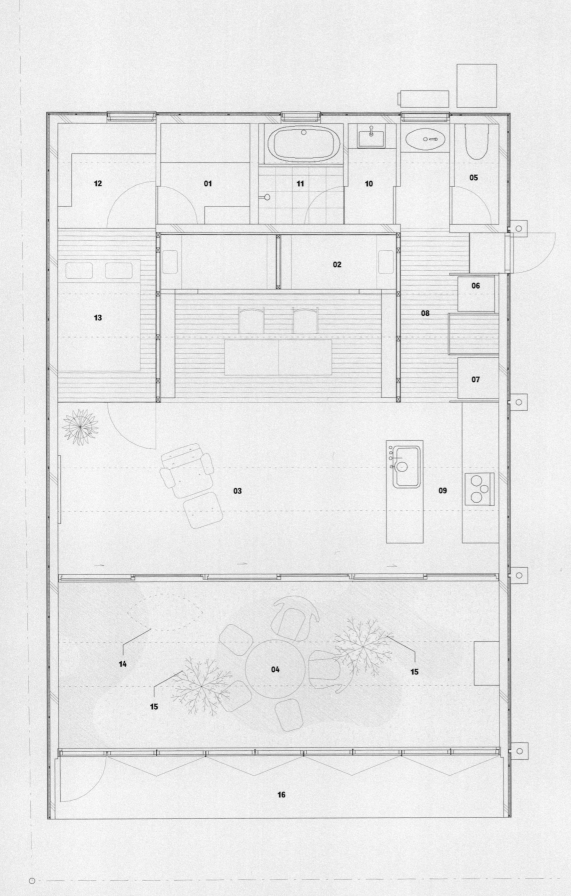

01 Closet
02 Children's room
03 Living
04 Dining
05 Half bath
06 Wash
07 Fridge
08 Utility
09 Kitchen
10 Washroom
11 Bathroom
12 Study room
13 Bedroom
14 Hammock
15 Olive
16 Porch

Floor plan

dIONISO LAB
HOUSE 77
Póvoa de Varzim, Portugal, 2010

family legacy. Inherited and transmitted across generations, this graphic language continues to evolve with new combinations. Careful not to detract from this historical homage, the building's street number behaves as a type of camouflage, subtly woven into the symbolic tapestry.

The metallic shutters can fold open or close shut depending on the level of privacy desired within. When closed, these shutters act as a form of urban

■ On a tight residential plot in a city by the sea, this narrow house discreetly catches the eye. The house sits in a vibrant area known for its fishing. This great cultural richness directly stimulates the approach to the project. An opportunity to revitalize some of the city's memories, the striking residence participates in the panoply of colors and materials that characterize the street.

CLAD IN A PANELED METAL SCREEN, THE THREE-STORY STRUCTURE'S MAIN FAÇADE FEATURES A NETWORK OF INTRIGUING GRAPHIC SYMBOLS ACROSS ITS EXTERIOR. Establishing the home's unforgettable identity, this eastern face incorporates perforated "siglas poveiras" onto its outer skin. These symbols, a proto-writing system developed in this Portuguese community, stand as a historical communication technique to mark personal and fishing related belongings. A sign of pride and lineage, the primitive writing represents an important part of one's

wallpaper, transmitting their energetic graphics into the neighborhood. These cutout graphics also illuminate the interior in a dazzling array of glowing symbols. Transmitting sunlight into the house during the day, at night this same system acts as a giant

three levels of the house maintain direct access to the outdoors. While large entry panels open up to the street, the top levels each include their own distinct balconies. These transitional spaces can either integrate into the interior layout when the window shutters are closed, or allow the residents to step out and observe the happenings of the neighborhood when fully opened. When open, sliding glass doors between the interior and the outer skin circulate fresh air into the house. Such a unique form of flexibility encourages a social lifestyle without impeding the private activities of those inside.

Numerous sustainable features appear throughout the residence, ranging from solar panels to plentiful natural light and ventilation. The house is also equipped with high thermal performance glass and aluminum frames. A small garden in the back embeds a sense of the natural into the urban context.

The simple house organizes itself in a vertical and hierarchical fashion. Social areas cluster on the lower floors while the private areas perch on the upper levels. To amplify visual impact and the dynamic interconnections between spaces, the interior divides into a series of half floors. A slender stairway, defined by the width of the plot, serves as the heart of the house. Painted in deep Yves Klein blue, one of the two side walls emphasizes the stairs' importance and continuity through the spaces. This high contrast area of rich blue, light wood, and bright white creates a compelling visual and spatial composition.

Facing west, the rear façade uses a system of aluminum venetian blinds. These blinds not only defend the interior from solar gains but also open the house onto an intimate back garden. Featuring floor-to-ceiling glazing on all three levels, this more private side of the residence relishes these unobscured site lines away from the public eye. The back side of the house angles out ever so slightly to introduce another level of protection against the sun. Supporting a strong connection to the outdoors, the open kitchen and living space bleeds out onto an exterior terrace. This dark wood terrace steps down on one side to reach a humble patch of grass below. Keeping in line with the outdoor

lantern, filling the street with warm light. The underlying flexibility of this façade enables the house to respond to both site conditions and the needs of its residents. As each panel opens up or closes shut, the house takes on a new and evolving appearance. All

When closed, these shutters act as a form of urban wallpaper, transmitting their energetic graphics into the neighborhood.

stairs, a series of square steppingstones lead the residents out into their personal green space—an urban refuge.

In spite of engaging different materials and aesthetic sensibilities, the house still manages to nurture a constructive and humble dialogue with its neighbors. **THE RESIDENCE, FRAMED BY STRUCTURES BOTH OLD AND NEW, CONTRIBUTES TO THE CHARMING CANVAS OF THE STREET.** This urban patchwork of brick, tile, steel, and concrete works together to define an overarching atmosphere for the area. Historically sensitive and refreshingly contemporary, the house, located in the center of Bairro Norte, shares some of the city's most precious memories. Just a short distance from the sea, the residence references the unique culture of the local population, revitalizing a legacy that has progressively been forgotten and abandoned. In this way, the house becomes a symbol in itself—a quiet confession of city pride. ∎

PLANS & DRAWINGS

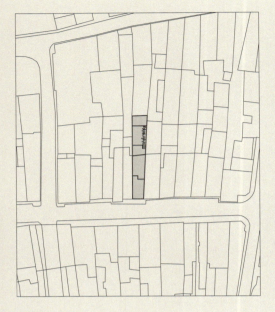

Site plan

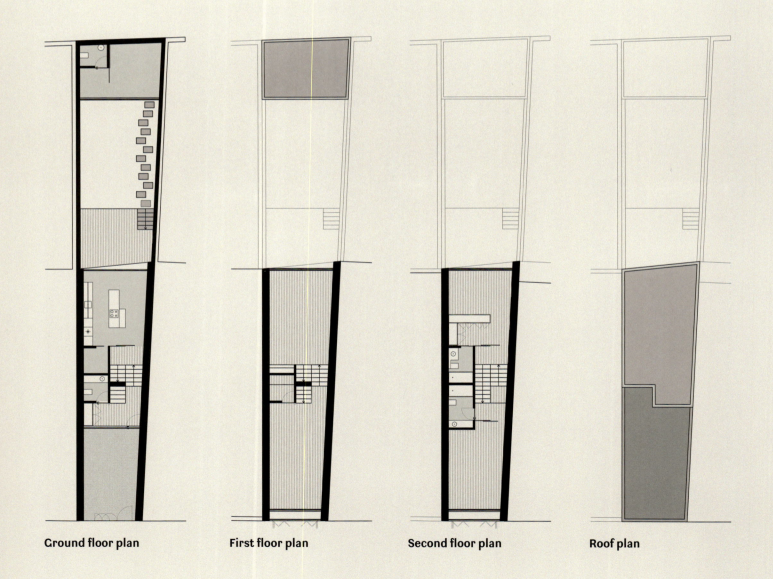

Ground floor plan   First floor plan   Second floor plan   Roof plan

East elevation

Spatial Strategy

This narrow house utilizes a system of folding screens to open and close the façade, depending on the needs of the family within. A floating stairs links all three levels, and the back side of the house opens onto an intimate garden.

●

Number of square meters:
232
Number of rooms:
2
Number of floors:
3
Number of residents:
2
Sustainable features:
High thermal performance glass,
solar panels,
natural light and ventilation
Outdoor areas/rooftop access:
Garden
Context:
City center
Type:
New construction

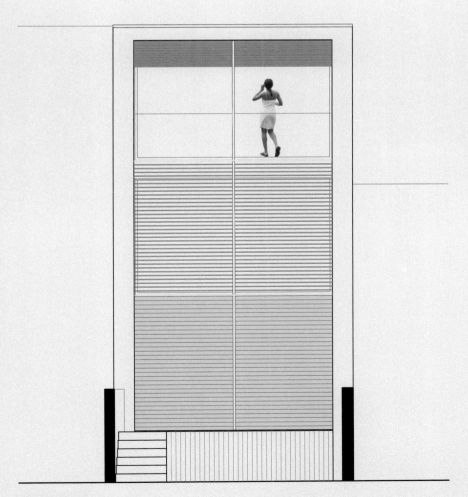

West elevation

Hiroshi Nakamura & NAP
HOUSE SH
Tokyo, Japan, 2005

■ **A**mysterious, bulging white monolith appears on a residential street in Tokyo. This highly unusual residence houses a young couple and their daughter. Set on a narrow site with close buildings on either side, the residential obelisk rises up three levels while hiding one underground floor below the street. **A BUILDING WITH MANY SECRETS, THE ELUSIVE HOUSE BEHAVES AS AN OPAQUE AND PROTECTIVE COCOON THAT NURTURES THE FAMILY INSIDE.**

The house retreats a few steps back from the street. Providing just enough space for the family's bikes or car, and the daughter to skip rope, the paved patch of outdoors acts as a transitional threshold between the building and the street. After raising the volume to the highest point possible, the building envelope then pushes out horizontally to the maximum building coverage. This gradual horizontal movement generates a distinctive character to the outside world while strategically widening the living room on the other side.

From the outside, the residence presents few clues about what lies within. The windowless façade swells toward the center, throwing soft shadows over the clean white surface. Intriguing all who pass by with its organic curving form, only the lucky family and their chosen guests know the interior implications of such a suggestive exterior treatment. The resulting interior concave pit takes on the shape of a giant kangaroo's pouch. Perfect for sitting in, lying on, or playing with, the enticing curvature of the softly rounded bench becomes a valuable and social fixture for the family. Encouraging the family to physically communicate with the wall, the supple surfaces beckon for attention. **THE REPEATED EXCHANGES BETWEEN THE FAMILY AND THE ARCHITECTURE PRODUCE A LIVELY AND EVOLVING RELATIONSHIP TO THE SPACE.** A carefully placed skylight above adds to the supple wall's allure, illuminating it under a perfect, even spotlight. This double-height intervention serves as the unconventional heart of the house.

The power of this iconic kangaroo pocket stems in large part from its multipurpose nature. Not only a place for lounging and gathering, the curving lines of the surface calibrate to bounce light into all corners of the house. Through the single-effective aperture, every level of the home retains access to the precious rays of the sun. All floor plates pull away from the wall to let the light travel down into the four levels of the house. This design feat bounces the sunlight entering through the skylight above off

the sculptural main wall, deflecting it into the living and dining spaces. From there, a glass panel in the floor meets the wall and brings more of its reflected light down into the ground and basement levels. A final rectangle fitted with a pane of pink glass fills the basement corridor with playful rose-colored light.

This decision to open up a part of each floor plate with glass not only benefits the transmission of daylight from floor to floor but also promotes visual vertical connections between spaces and levels. Through these apertures, the parents can observe their daughter playing on the sculpted wall from one and even two floors down. This embedded transparency makes the parents' task of keeping an eye on their child much easier. Simultaneously, these visual links give the daughter a sense of security, as she can easily locate her parents throughout the house while still having the freedom and space to feed her imagination.

Hidden just off the street, the main entrance remains completely out of sight. A refined fence made of a light wood leads guests down a narrow exterior corridor to the front door. Once inside, a white spiral staircase stands as the connective tissue across all floors. The treads of these white stairs are perforated with an array of circular cutouts. This network of holes overlap into a layered constellation for light to pass through—a swirling path of luminous stars. With the children's rooms situated at the very top of the residence, the level below accommodates the kitchen, dining, and living spaces, complete with the nurturing seating nook. The main entry level one floor lower holds an extra bedroom and bath. A final basement floor culminates in the couple's master bedroom and a tranquil reading room.

A perfect home for kids and adults alike, the design approach of this project exudes a maternal quality—a refuge from the metropolis. The glowing residence solves the riddle of how to make a light house in a dark site. Windowless and wonderful, the top-lit interior spaces instigate a revitalizing approach towards family life. Soft bulging corners exude a feminine mystique that cultivates a healthy relationship between the couple, their daughter, the architecture, and the city. The child-oriented spaces keep the family young at heart. Brimming with dappled light and velvety tactility, this impressive project raises the bar for the urban dwelling by tapping into the senses. By working with the imagination, and defying expectations, a grand incubator for creativity emerges. ∎

From the outside, the residence presents few clues about what lies within.

226 <u>House SH</u>

PLANS & DRAWINGS

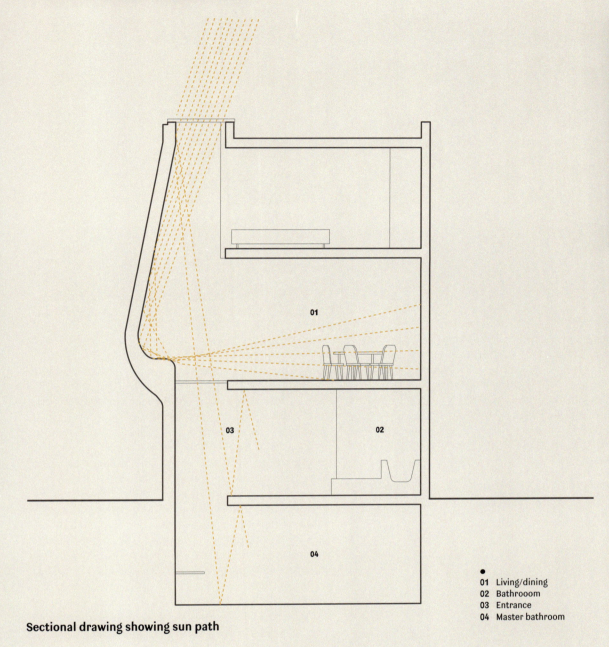

Spatial Strategy

A windowless house channels light from above. The sculpted, multipurpose bulge set into the front façade transmits daylight down through the different levels of the interior via its concave surface.

●
Number of square meters:
87
Number of rooms:
3
Number of floors:
4
Number of residents:
2 adults, 1 child
Sustainable features:
N/A
Outdoor areas/rooftop access:
N/A
Context:
City center
Type:
New construction

Sectional drawing showing sun path

●
01 Living/dining
02 Bathrooom
03 Entrance
04 Master bathroom

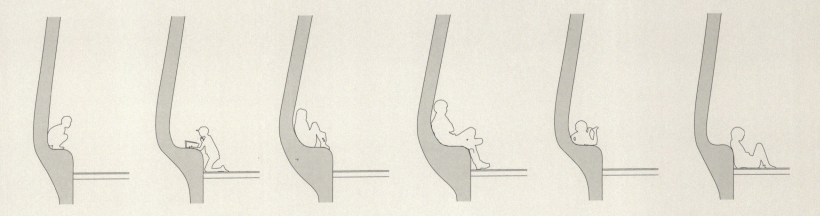

Diagram of uses of main wall cavity

Studio Velocity
HOUSE IN CHIHARADA
Okazaki, Japan, 2012

■ This refreshing approach to living consists of a renovated two-story main house split in half and a new house for a young couple built close by. In spite of the spacious, low-density surroundings, the two structures nestle close to one another. To avoid facing each other, a rounded volume serves as the ideal shape for the new residence. This circular form sits against the corner of the square-shaped volume of the main house. The arrangement creates a valley-like space in between the two buildings that spreads open as it reaches toward the outside.

THE ROUND AND TALL HOUSE RESTS ON AN IRREGULAR SHAPED SITE, ACCENTUATING THE GEOMETRIC QUALITIES OF THE SETTING. A layout of eight rooms includes an open living, dining, and kitchen area on the top floor. Linked to gardens on all sides, the ground floor incorporates a master bedroom, a children's room and play area, laundry facilities, and a bathroom with a separate wash area. Using a circular floor plan, the building influences and shapes the varied geometric gardens just outside. These gardens can also be shared with the main house. Each room on the first floor of the round building integrates a door that opens to the gardens. This entry level comprises a number of small rooms and a bathroom. On the second floor, a single large hall behaves as a communal gathering place. Downstairs and upstairs remain relatively close together thanks to the lowered height of the second floor slab. The slab, topped with light wood flooring, lies between the two levels. This particular placement facilitates visual access to the garden grounds below. These gardens stay in view regardless of where one stands inside the house. From the center of the second floor to the enclosed staircases and downstairs rooms, all parts of the house maintain a connection to the outdoors and nearby nature.

A slender exterior staircase winds its way up to the second level of the residence. When passing through this entrance, guests then encounter an array of closed staircases arranged around the living room. These slender structures of shifting heights form a striking array against the sunlight-filled, high-ceilinged living area. While keeping with the white color palette on the exterior, the interior instead transports the family into a world of dark, rich woods. **SUCH CONTRAST ADDS TO THE VISCERAL IMPACT OF THE STAIRS AS PORTALS BETWEEN TWO DIFFERENT BUT INTERCONNECTED WORLDS. EACH OF THE FOUR ENCLOSED STAIRCASES CONNECTS TO AN INDIVIDUAL ROOM ON THE FIRST FLOOR.** The boxed-in

stairs retain an opening at the very top. Bringing light and views into the lower level, these apertures nearly reach the roof. A viewport to the sky, the children on the ground floor can look up through the stairwell and admire the clouds passing by and the natural light that pours down. This ingeniously simple solution for circulating people and light enables the family to remain always conscious of both ground and sky throughout the multiple levels of the building.

Evoking the feeling of a street in a small town, the densely programmed living space elevates the public and social spaces up a level. A family dining table sits at the very center of the space. Bordered by the enclosed stairs on every side, this dining area offers a feeling of privacy in spite of its public context. By inverting the typical two-story layout, the private bedrooms below can enjoy the intimate gardens while the social level above features numerous views across all directions of the site. These curated vistas frame glimpses of

a brilliant green hill overlooking the property and verdant scenery to the west. The size, shape, and location of each window and skylight corresponds to carefully considered views of the outside. Plants and small trees infiltrate the interior, softening any hard edges in their path. Leaves and splashes of green further erode distinctions between indoor and outdoor living. **THIS CLEVER LAYOUT ELIMINATES THE DISCONTINUITY THAT USUALLY EXISTS BETWEEN FLOORS. INTERRELATING EACH AREA, THE HOUSE UNFOLDS AND UNITES INTO A CONTINUOUS NEW LIVING ENVIRONMENT.** The equal engagement and intersection of both indoor and outdoor spaces from upstairs to downstairs dissolves the hierarchy between the first and second floors. Now, individual functions and sceneries fluidly mix together into a harmonious residential lifestyle.

The design meets the highest energy saving ranking system in Japan. High performance glass and thermal insulation work together to generate an extremely energy efficient home. Improving both the indoor environment and the residence's relationship to sustainability, the building's frame, furniture, and finish materials are all chosen with longevity in mind.

Iconic without being pretentious, the whimsical home captures a modern Japanese sensibility. The close relationship between the old and new structures builds a vibrant, miniature village within an otherwise open and low density country setting. This gracious approach mixes modest comforts with air, light, and nature. ∎

PLANS & DRAWINGS

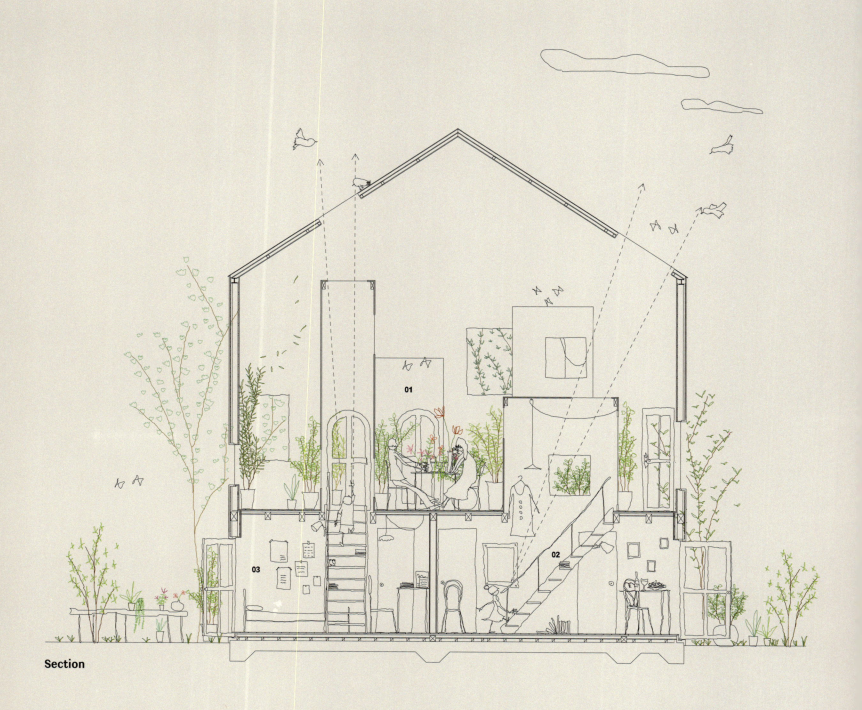

Section

●
01 Living/dining/kitchen
02 Washroom
03 Children's room

HOUSE IN CHIHARADA

Ground floor plan

Spatial Strategy

Two independent but related houses sit in close proximity to one another. The rounded new house features garden access on the lower level. The unusual house inverts the public and private spaces so the bedrooms can open up directly to the outdoors.

● **Number of square meters:**
110.56
Number of rooms:
8
Number of floors:
2
Number of residents:
2 adults, 1 child
Sustainable features:
Efficient insulation and glazing
Outdoor areas/rooftop access:
Garden
Context:
Outskirts
Type:
New construction

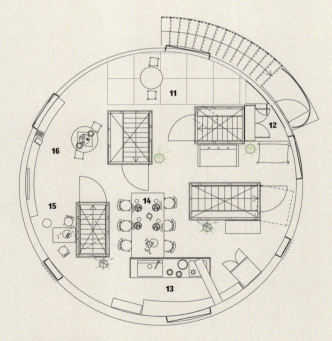

First floor plan

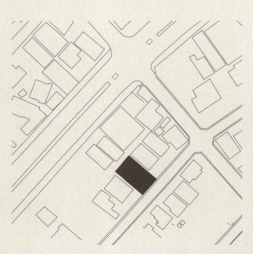

Site plan

● 02 Washroom
03 Children's room
04 Main house
05 Bench
06 Laundry drying area
07 Bathroom
08 Garden table
09 Counter-/wash basin
10 Master bedroom
11 Tatami room
12 Entrance
13 Kitchen
14 Dining
15 Children's space
16 Living area

Kimihiko Okada
TODA HOUSE
Hiroshima, Japan, 2011

■ On a gentle perch in a residential area of Hiroshima, this sculptural house overlooks a vast view of the Inland Sea and the island of Miyajima. The land of this area is developed into platforms with multiple levels. Several of the client's wishes define this home's unique appearance. The first consideration deals with how to capture a view over the roofline of the neighboring house that stands one level lower. A second parameter focuses on security. With the site located at the edge of the residential area, safety becomes an important criteria. Thirdly, the home requires a degree of spatial flexibility to accommodate the possibility of a future extension if the client opens a small shop on the premises. The outcome of these individual requests lifts the house a level up off the ground. Like a bird's nest, the floating residence embodies architecture's primary function as a shelter from disturbance. The house proves both open to the inspiring views of the area and efficiently protected from the unpredictable nature of the environment.

Slab and roof consist of a single continuous plate. The extended plate not only makes it possible to complete the future extension but also softens the building's overall impression from the ground level. A strategically placed, penetrating staircase generates variations of circulation and diverse spatial relationships. Spandrel walls shift in height according to the thickness of the slab. Together, the slab and spandrel walls choreograph a continuous yet individuated environmental experience.

A main opening at the heart of the house catches the breeze that passes through the structure's slender stilts and transfers it up and into the raised home. Stacked ventilation grants the spiral house a refreshing form of natural air conditioning. Due to

the elongated shape and the raised structure, the ground-level garden provides an open and bright landscape. **THIS CENTRAL PLACEMENT OF THE PERSONAL GREEN SPACE ADDS A NEW PARK TO THE NEIGHBORHOOD.** The elevated structure above also stimulates an optimal climate for the garden below. With plentiful access to sunlight and cool breezes, this unexpected oasis becomes a rich landscape for plants to grow.

THE FLOOR PLAN CIRCUMSCRIBES THE DOWNSTAIRS GARDEN LEVEL. Shaped into a square with rounded edges, the center of this geometric layout introduces a sinuous void. This multipurpose void transforms a generic square layout into a dynamic, fluid interior path that provokes a sensation of

With plentiful access to sunlight and cool breezes, this unexpected oasis becomes a rich landscape for plants to grow.

urban environment. The residence's rare appeal stems from its commitment to doing more with less.

Wraparound glazing on both sides of the interior follow its curving path. These generous swatches of glazing produce a spectacular 360° panorama of the area. Whether admiring the rich greens of the surrounding forest or gazing out past the tiled rooftops towards the bay, this enthralling home doubles as an exclusive lookout point. Flooded with daylight, the continuous windows also grant the occupants the luxury of tracking the course of the sun throughout the day.

Silver legs that support the house extend up into the interior. These legs become rounded columns once inside, rhythmically introducing separations between the endless glazing. The highest point of the house eventually swirls back over itself, culminating in a rooftop terrace that faces onto the main street. A second larger terrace sits one level below near the main entry to the house. Bordered by the entrance on one side and the dining room on the other, this indoor/outdoor space promotes a social atmosphere for open-air dining or lively gatherings.

A cozy living room resides at the highest point of the layout covered in a light herringbone flooring. This patterned flooring gives the luminous space a rhythmic and energetic quality. Softened with a brightly patterned rug and potted plants, the contemplative room affords its residents the unique experience of relaxing on top of the world.

movement. The resulting interior experience develops a contemporary interpretation of the classic railroad style home—where each room bleeds into the next. After ascending from the outdoor ground-floor staircase, the interior spaces begin to wind across the site over two levels. Such a gradual but constant upward movement exalts the interior atmosphere.

WHAT MAKES THIS HOUSE SO UNUSUAL LIES IN ITS INTEREST OF QUALITY OVER QUANTITY. Rather than maximizing the amount of usable floorspace, the interior approach curates a modest yet noble footprint. The majority of the site remains outdoors and wild, a testament to the value of the natural within an

The cheerfully landscaped garden behaves as both a public and private part of the house. As the residence protectively circles overhead, strong sight lines link the indoor and outdoor spaces. This forging of vertical and horizontal connections between both inside and out allows the different family members to relate and respond to one another across multiple levels. A humble refuge in the public domain bridges private needs with social connection through an exuberant and uplifting architectural expression. ∎

PLANS & DRAWINGS

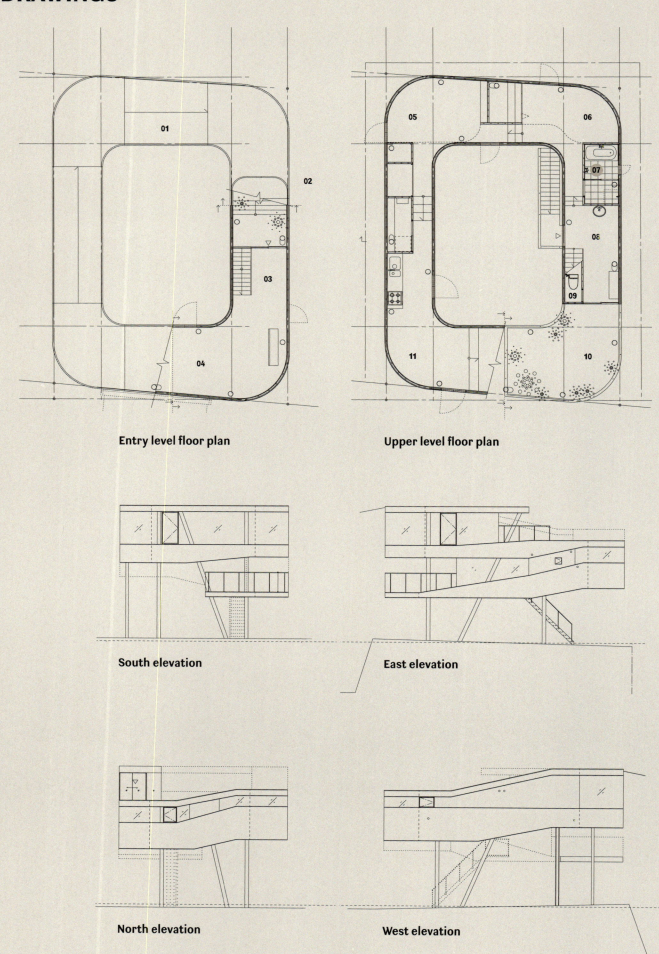

Entry level floor plan

Upper level floor plan

South elevation

East elevation

01 Roof
02 Roof terrace
03 Study
04 Living room
05 Bedroom 1
06 Bedroom 2
07 Bathroom
08 Entrance
09 Half bath
10 Terrace
11 Dining room

North elevation

West elevation

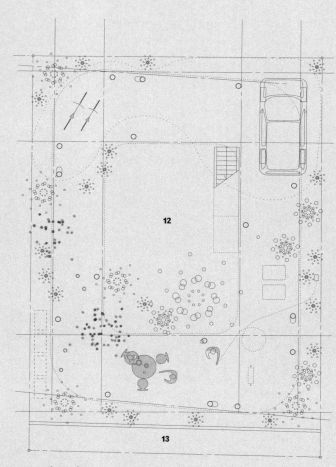

Garden plan

12 Parking
13 Retaining wall

Kimihiko Okada
TODA HOUSE

Spatial Strategy

Floating on stilts above a ground floor garden, this swirling home rethinks the continuous railroad style floor plan. With each space bleeding into the next, all rooms connect to one another, the dramatic vistas, and the garden below.

●

Number of square meters:
114.26
Number of rooms:
5
Number of floors:
3
Number of residents:
3
Sustainable features:
Natural ventilation
Outdoor areas/rooftop access:
Courtyard and terrace
Context:
Residential area
Type:
New construction

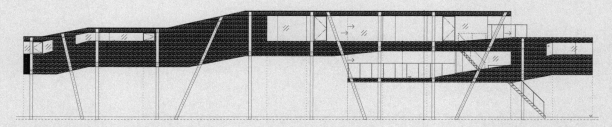

Exploded elevation/outside wall

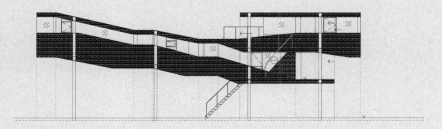

Exploded elevation/inside wall

Clare Cousins Architects
BRICK HOUSE
Prahran, Victoria, Australia, 2011

lot. Solving this privacy issue for the house and garden stands as the primary consideration influencing the remodel.

DESIGNED FOR A YOUNG FAMILY OF FOUR, THE SIX-ROOM HOUSE CATERS TO THEIR CURRENT AND FUTURE NEEDS. Reminiscent of the classic homes developed in Los Angeles for the Case Study House Program during the 1940s, 50s, and 60s, the residence sets up a captivating relationship between indoor and outdoor living. Whitewashed interiors flooded with light create a serene mood for appreciating the nuances of nature while optimizing feelings of privacy and comfort. A flexible space, referred to as the studio, rests atop the new garage at the rear of the site. This new double-story mass helps screen out the large neighboring buildings. The single-story addition to the house engages with the garden through a meandering, fluid façade of glass and brick.

Working with a long and narrow site, the rear façade manifests as a continuous white brick surface. The exterior of both structures revolves around the luminous pool and courtyard. Soft internal and external corners maximize the glazed vista into this garden area. A large built-in window seat and sofa in a charming shade of marmalade follows the dynamic window line in the living room. This welcoming seating arrangement provides immersive views of the courtyard while simultaneously activating the interior. Situated around a modern fireplace with white brickwork and clean lines, the warm furniture accents of the living room make for an intimate way of relating to the space. Juxtaposing with the quiet and regular lines of the fireplace, potted plants and a circular rug with deep red-toned patterning break up the formal language of the architecture.

■ This gracefully undulating project consists of an addition and renovation to an Edwardian house in a Melbourne suburb. Craving privacy from looming neighboring flats, two new structures cocoon a private central courtyard space and pool between them. The existing Edwardian home sits comfortably on a street lined with single-story period cottages. At the rear of the site, this density escalates into a cluster of three-story townhouses that directly overlook the private open spaces of the

The insertion of a small light court at the end of the extended original hallway blurs the boundary between the interior and garden experiences. This subtle transition between old and new spaces appears at the junction of the two hallways. A shift in light quality and an inversion of material usage further accentuate this transitional area.

The insertion of a small light court at the end of the extended original hallway blurs the boundary between the interior and garden experiences.

issues. Currently used as a guest bedroom, the flexible studio can accommodate a variety of uses.

In addition to its elegant charm, the house also introduces a number of sustainable features into its design. The main windows of the extension orient to the north and are protected by a large eave that pulls away from the façade to shade the glass from solar heat gain. Although all glass is double glazed and low-E-coated, the west-facing windows of the studio are further shielded by the vertical timber battens. The house integrates two underground 5,000-liter storage tanks. The first tank collects rain water to water the vegetable garden and top up the small plunge pool. A second container stores treated grey water to irrigate the garden and flush toilets. Light woods and black and white tiles are just a few of the robust and readily available materials selected for the interior and exterior. These classic materials require minimal maintenance and age well over time. A private oasis in a very public neighborhood, this residence cultivates a timeless ambiance for living well. ∎

Serving as the new heart of the house, the centrally located kitchen resides in the new extension at the end of the hallway. In spite of being tightly planned,

the airy layout maximizes storage and functionality to suit family life. Mixing high levels of transparency and opacity, the charcoal-glazed brick kitchen volume forms a buffer for the playroom just behind. The dark brick accents of this central core ensures a tasteful composition that continues to catch the eye. Without the need for doors or fixed partitions, the floor plan achieves clear distinctions between spaces. A generous study area located next to the dining table and kitchen features a perforated screen that can be drawn when concentrated work must be done. This layout keeps everything open and in close proximity to one another, while carving out nooks for retreat.

The bright and streamlined kitchen looks out onto the garage building with its second floor studio. Casting varying shadow patterns across the courtyard during the day, this omnipresent studio incorporates an undulating timber batten façade that relates to the curving glass walls of the main house. The light, billowy nature of the wooden screen allows the studio windows behind to swing open and naturally ventilate the space. As well as promoting ventilation, this striated façade offers different levels of transparency to compensate for the site's privacy

PLANS & DRAWINGS

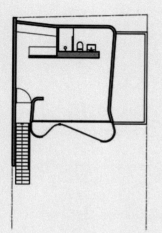

First floor plan

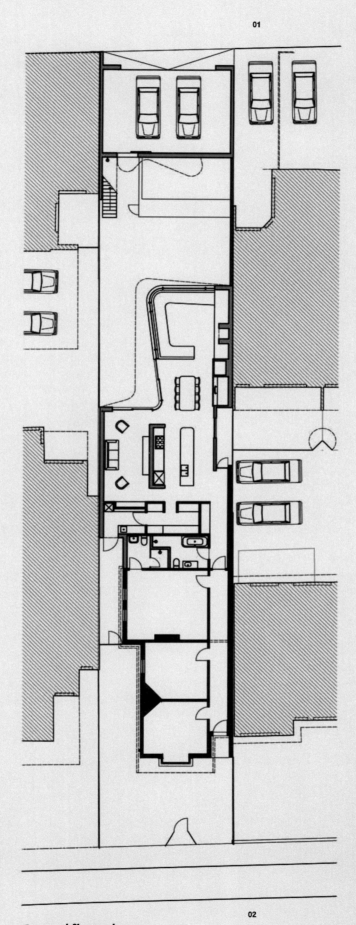

01

Ground floor plan

02

Spatial Strategy

A sinuous renovation and extension orients around a charming courtyard space. Maximizing privacy, the renovated original home hosts the main living quarters while the addition introduces a multipurpose studio and guest area.

●
Number of square meters:
205
Number of rooms:
6
Number of floors:
2
Number of residents:
4
Sustainable features:
Repurposed materials, north-facing windows with eave to protect from solar heat gain, double-glazed, low-E-coated glass, timber batten sunscreens, natural ventilation
Outdoor areas/rooftop access:
Courtyard
Context:
City center
Type:
Addition and renovation

●
01 Lane
02 Street

Grupo Aranea
CASA LUDE
Cehegin, Spain, 2011

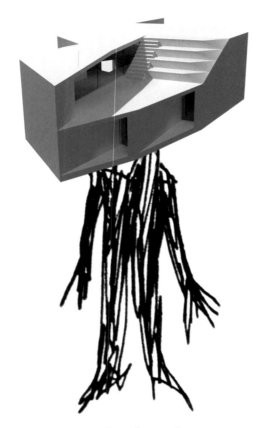

environment. Due to the density of the area, the residence never opens directly to the nearby buildings. All apertures look into the narrow streets and alleyways. From here, the family enjoys private vistas of the landscape which include the Burete mountains and the Hill of Saint Augustine.

In spite of opting for a more iconic and futuristic aesthetic compared to the other buildings in the neighborhood, the house still shares the area's same compact and introverted characteristics. The unconventional interior layout stages a one-of-a-kind, continuous, and cinematic space filled with music. These intricate and interconnected floor plans move across different levels of various heights and directions. This fluid arrangement of spaces extends to the roof—a natural continuation of interior life. Within this complex space full of light, many of the secrets to its appeal stay hidden.

PANORAMIC 360° VIEWS MOTIVATE THE HOUSE TO TAKE ADVANTAGE OF ITS NEW FOUND HEIGHT VIA A SERIES OF TERRACES. These terraces face in multiple directions, extending the minimalist interior out into the open air. An ideal spot for passing the afternoon playing the guitar, the main rooftop deck folds into a stepped amphitheater. Potted plants sprinkle over the stepped terrace, presenting the pristine white space with a touch of rich color. These steps lead up to an even higher rooftop deck where the gleaming white planes dissolve into blue skies. A thin protective railing ensures unobstructed views of the city and rolling green hills beyond. Floor-to-ceiling glazing accompanied by sliding glass doors craft a seamless transition between the interior and exterior. This dramatic mix of indoor and outdoor spaces encourages the family to forge a meaningful connection with the city.

The angular interior facilitates diverse sight lines and relationships across spaces. These sloping, gleaming surfaces add depth to the interior, highlighting certain corners and walls while throwing others into shadow. Vertically oriented rectangular openings offer glimpses of the traditionally tiled

■   Leveraging the roof over the mother's family house, this new residence transforms a traditionally unused site into a breathtaking additional plot for the son. The resulting contemporary rooftop dwelling towers over the more historical Spanish neighborhood. Making the most of its privileged situation, the elevated house relates to its environment in a particular way. Built on a plinth of history, the striking addition embodies a generational palimpsest of a single family. This layering of generations also applies to the architecture. The humble, weathered original structure in all its rugged honesty contrasts with the sleek, crisp white lines of the modern structure above.

The house has no windows. Instead, the striking residence introduces a series of open patios. These patios appear on every level of the home and give the layout a particular manner of relating to the

Spanish rooftops. These curated views become abstract canvases—colorful swatches of city life layered over decades. With these timeless views behaving as the interior's only connection to ornament, the rest of the spaces feature clean white walls. These efficient, bright surfaces produce a certain flexibility of use. Walls and ceilings double as backdrops for the projection of films, video clips, and other selected imagery.

design approach maintains a considerate attitude and sense of respect in regard to the characteristics present on this particular site. One instance of how the home responds to its context relates to its final geometry. The design works within the parameters imposed by the original load bearing walls of the existing home. These structural supports clearly influence the final outcome, building footprint, and formal agenda of the project.

Every aspect of the house serves as a lesson for how to optimize available, and at times limited, resources. Intensifying the experience of daily life in the city center, the residence also remains respectful of the original homes of the mother and her sister. The light house functions as its own unique and autonomous universe. **RESEMBLING AN ANGULAR UFO DROPPED FROM THE SKY, THE COMPELLING BUILDING CARVES OUT ITS OWN PERSONAL ATMOSPHERE SEPARATE FROM THE WORLD BELOW.** The high contrast residential monolith achieves a dramatic juxtaposition as it surreally rises up from the somewhat stagnant neighborhood. At the same time, the

The complexity of this type of construction derives from its commitment towards engaging with the city's urban center. Rather than expanding the periphery and destroying more of the region's precious farm land, the house instead thrives within the existing context. The challenge to build on top of an existing structure results in an architectural intervention with both grace and foresight. By building up instead of building out, the residence not only fights against the tendency towards urban sprawl but also initiates an intriguing dialogue between past, present, and future. ∎

PLANS & DRAWINGS

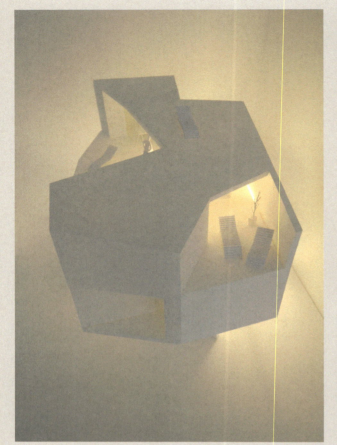

Model study

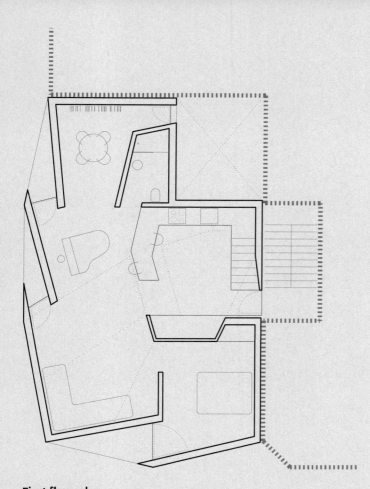

First floor plan

CASA LUDE

Spatial Strategy

Perched on top of the original family home, this new futuristic residence winds its way through a series of outdoor spaces. These terraces take the place of traditional windows, allowing the interior to engage with the outdoors in a meaningful way.

●

Number of square meters:
224
Number of rooms:
1
Number of floors:
3
Number of residents:
2
Sustainable features:
N/A
Outdoor areas/rooftop access:
Rooftop terrace and courtyards
Context:
City center
Type:
Addition

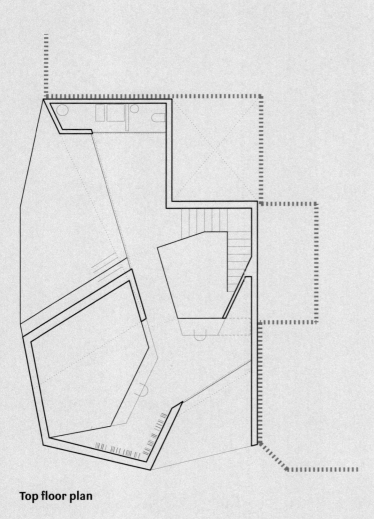

Top floor plan

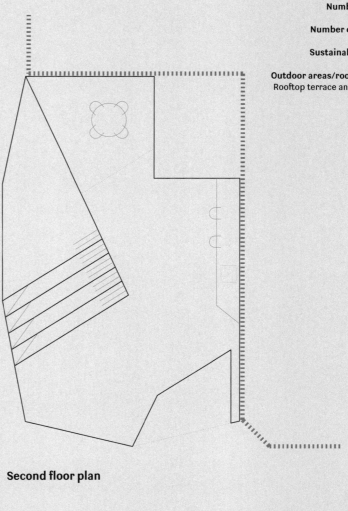

Second floor plan

INDEX

A

A21 STUDIO
Vietnam
www.a21studio.com.vn

The Nest
Photography: Hiroyuki Oki
(pp. 056–059, 060 right
bottom, 161); a21studio
(pp. 060 left top)
pp. 056–063

ALPHA-VILLE
Japan
www.a-ville.net

New Kyoto Town House
Architect: Kentaro Takeguchi,
Asako Yamamoto/Alphaville
Additional credits: Kazuo
Takeguchi (structural
engineer); Kawana Kohgyo
(general contractor)
Photography: Kei Sugino,
Kentaro Takeguchi
pp. 188–193

ANDREW MAYNARD ARCHITECTS
Australia
www.maynardarchitects.com

House House
Additional credits: Skyrange
Engineering (external,
internal and red cedar
cladding, Victorian ash and
spotted gum veneer, window
frames); Shadefactor
(motorized and fixed
aluminum louver shading);
Big River Timbers (flooring
spotted gum veneer)
Photography: Peter Bennetts,
www.peterbennetts.com
pp. 020–027

Vader House
Photography: Peter Bennetts,
www.peterbennetts.com
pp. 090–095

AREAL ARCHITECTEN
Belgium
www.arealarchitecten.eu

House DV Wilrijk
Architect: Chris Eeraerts,
Thomas Cols/Areal
Architecten
Additional credits: Dakwerken
Costermans (contractor
façade and roof)
Photography: Chris Eeraerts/
Areal Architecten
pp. 124–129

B

BFS D
Germany
www.bfs-design.com

Gas Station
Architect: Stefan Flachsbarth,
Michael Schultz/bfs d;
Thomas Brakel/planbb
Additional credits: M+N
Ingenieure (static engineer)
Photography: Annette Kisling
pp. 168–173

BLACK LINE ONE X ARCHITECTURE STUDIO
Australia
www.bloxas.com

Profile House
Additional credits: Genjusho
(builder); Clive Steele
Partners (engineer);
Wilsmore Nelson Group
(building surveyor)
Photography: Peter Bennetts,
www.peterbennetts.com
pp. 096–103

BRANDT + SIMON ARCHITEKTEN
Germany
www.brandtundsimon.de

Schuppen
Additional credits: Ingenieur-
büro für Tragwerksplanung
Dr.-Ing. Christian Müller,
Frank Niehues (planning of
structural framework)
Photography: Michael Nast,
www.lichtkombinat.de
(pp. 149 right bottom, 149
left, 150); Brandt+Simon
Architekten (pp. 148, 149
right top, 151)
pp. 148–153

BRAUN UND GÜTH
Germany
www.braun-gueth.com

Pünktchen
Architect: Sascha Christine
Braun, Daniel Güth/Braun
und Güth Architekten
Additional credits: S. S.
Rothenberger (concept);
S. S. Rothenberger,
K. Mainka/E15, S. Braun
(interior design); DYNAMO
Studio, D. Güth, M. Scherer/
HALLO WELT (façade design);
Daniel Kaufmann/DK
Steintechnik (stonemason);
M. Graupner/Atelier n.4
(carpenter); C. Köhler/
Köhler Haustechnik
(technical installation);
M. Scherer/HALLO WELT
(graphic design)
Photography: Peter Unsinn,
www.peterunsinn.com
pp. 154–159

C

CASE-REAL
Japan
www.casereal.com

White Dormitory for IL VENTO
Architect: Koichi Futatsu-
mata/CASE-REAL
Additional credits: Tsuyoshi
Matsuzawa (direction)
Photography: Hiroshi Mizusaki
www.loop-pc.jp (pp. 064
middle+bottom, 065–067);
CASE-REAL (pp. 064
top+middle)
pp. 064–069

C. FISCHER INNENAR-CHITEK-TEN
Germany
www.fischerinnen.de

Townhouse
Photography: Florian
Kleinefenn, www.kleinefenn.
com (pp. 160–163,
165–167); Pia Ulin (p. 164)
pp. 160–167

CLARE COUSINS ARCHI-TECTS
Australia
www.clarecousins.com.au

Brick House
Additional credits: Clare
Cousins, Michael McManus,
Sarah Lake, Renae Brown,
Tara Ward (project team);
Maben Group (builder);
Eckersley Garden
Architecture (landscape
design)
Photography: Shannon McGrath,
www.shannonmcgrath.com
pp. 242–247

D

DIONISO LAB
Portugal
www.dionisolab.com

House 77
Architect: José Cadilhe/
dIONISO LAB
Photography: FG + SG
Fotografia de Arquitectura,
www.fernandoguerra.com
(pp. 212–213, 214 right,
215–217); dIONISO LAB
(pp. 214 left)
pp. 212–219

E

EASTERN DESIGN OFFICE
Japan
www.easterndesignoffice.com

Mon Factory/House
Additional credits: Hojo
Structure Research Institute
(structure planning)
Photography: Koichi Torimura
pp. 028–035

EDWARDS MOORE
Australia
www.edwardsmoore.com

Dolls House
Photography: Fraser Marsden,
www.frasermarsden.com
pp. 070–075

F

FRAN SILVESTRE ARQUI-TECTOS
Spain
www.fransilvestrenavarro.com

House on Mountainside Overlooked by Castle
Additional credits: Fran
Silvestre, María José Sáez
(principal in charge); Ángel
Ruiz (collaborator architect);
Pedro López (building
engineer); Alfaro Hofmann
(interior design)
Photography: Fernando Alda,
www.fernandoalda.com
pp. 110–117

G

GRUPO ARANEA
Spain
www.grupoaranea.net

Casa Lude
Architect: Francisco Leiva
Ivorra, Martín López Robles/
Grupo Aranea
Photography: Jesús Granada,
www.jesusgranada.com
(pp. 250–251); Francisco
Leiva (pp. 249); Grupo
Aranea (pp. 248, 252)
pp. 248–253

H

HIDEAKI TAKAYANAGI
Japan
www.hideakitakayanagi.jp

Life in Spiral
Photography: Takumi Ota,
www.phota.jp
pp. 174 – 179

HIROSHI NAKAMURA & NAP ARCHITECTS
Japan
www.nakam.info

House SH
Photography: Daici Ano,
www.fwdinc.jp (pp. 222, 223,
225, 226 top); Naoki Honjo
(pp. 220); Hiroshi Nakamura
& NAP (pp. 221, 224, 226
bottom)
pp. 220 – 227

Optical Glass House
Photography: Koji Fujii/
Nacasa and Partners,
www.nacasa.co.jp (pp. 042,
043 top+bottom, 044 – 047);
Hiroshi Nakamura & NAP
(pp. 043 middle)
pp. 042 – 049

J

JONATHAN TUCKEY DESIGN
United Kingdom
www.jonathantuckey.com

Shadow House
Photography: Dirk Lindner,
www.dirklindner.com
(pp. 142, 143, 144 bottom,
145); James Brittain
www.jamesbrittain.co.uk
(pp. 144 top)
pp. 142 – 147

J. MAYER H.
Germany
www.jmayerh.de

OLS House
Additional credits: AB Wiesler,
Michael Gruber (architect on
site); Gunter Kopp
(structural engineer);
IB Funk und Partner (service
engineers); Kurz & Fischer
(building physics)
Photography: David Franck,
www.davidfranck.de
pp. 194 – 199

K

KIMIHIKO OKADA
Japan
www.ookd.jp

Toda House
Additional credits: Structured
Environment (structural
engineering); System Design
Laboratory (mechanical
engineering)
Photography: Toshiyuki Yano
pp. 236 – 241

L

LEVEL ARCHITECTS
Japan
www.level-architects.com

Skate Park House
Architect: Kazuki Nakamura,
Kenichi Izuhara/LEVEL
Architects
Photography: Junji Kojima
www.45g.jp
pp. 076 – 083

LISCHER PARTNER ARCHITEKTEN
Switzerland
www.lischer-partner.ch

Stadtvillen
Photography: Roger Frei,
www.rogerfrei.com
pp. 130 – 135

M

MA-STYLE ARCHITECTS
Japan
www.ma-style.jp

Light Walls House
Architect: Atsushi Kawamoto,
Mayumi Kawamoto/mA-style
Architects
Photography: Kai Nakamura
pp. 200 – 205

MARC KOEHLER ARCHITECTS
Netherlands
www.marckoehler.nl

House Like Village
Additional credits: Made UP
interior works (furniture
co-design)
Photography: Marcel van der
Burg, www.primabeeld.nl
pp. 180 – 187

MEIXNER SCHLÜTER WENDT
Germany
www.meixner-schlueter-wendt.de

Residence Z
Photography: Christoph
Kraneburg, www.kraneburg.net
pp. 136 – 141

MOON HOON
Republic of Korea
www.moonhoon.com

Lollipop House
Architect: Moon Hoon,
Moonbalsso
Additional credits: Lee Ju Hee,
Kim Dong Won, Park Sang Eun
(project team)
Sketch: Moon Hoon (pp. 087)
Photography: Nam Goong Sun
pp. 084 – 089

S

SNARK
Japan
www.snark.cc

House in Keyaki
Architect: Sunao Koase/
SNARK, Shin Yokoo/OUVI
Photography: Ippei Shinzawa
www.ippeishinzawa.com
pp. 118 – 123

STUDIO VELOCITY
Japan
www.studiovelocity.jp

Montblanc House
Photography: Kentaro
Kurihara
pp. 008 – 013

House in Chiharada
Photography: Kentaro
Kurihara
pp. 228 – 235

SPACECUTTER
United States
www.spacecutter.com

Carved Duplex
Additional credits: GRANT
engineering
Photography: Michael
Vahrenwald, www.
michaelvahrenwald.com
pp. 014 – 019

T

TAKESHI HOSAKA ARCHITECTS
Japan
www.hosakatakeshi.com

Daylight House
Photography: Koji Fujii/
Nacasa & Partners,
www.nacasa.co.jp
pp. 036 – 041

Outside In
Photography: Koji Fujii/
Nacasa & Partners,
www.nacasa.co.jp
pp. 206 – 211

TATO ARCHITECTS
Japan
www.tat-o.com

House in Yamasaki
Architect: Yo Shimada/Tato
Architects
Photography: Kenichi Suzuki
pp. 104 – 109

TERRA E TUMA
Brazil
www.terraetuma.com.br

Maracanã House
Photography: Pedro Kok,
www.pedrokok.com
pp. 050 – 055

OUR HOUSE IN THE CITY

New Urban Homes and Architecture

This book was conceived, edited, and designed by Gestalten.

Edited by Sven Ehmann and Sofia Borges
Text and preface by Sofia Borges

Cover by Floyd E. Schulze
Cover photography by FG + SG Fotografia de
Arquitectura, "House 77" by dIONISO LAB
Art direction by Floyd E. Schulze
Layout by Regine Rack
Layout assistance by Pepita Köhler
Typeface: Ludwig by Fred Smeijers

Proofreading by Transparent Language Solutions
Printed by Livonia Print, Riga
Made in Europe

Published by Gestalten, Berlin 2014
ISBN 978-3-89955-518-9

For more information, please visit
www.gestalten.com.

Bibliographic information published by the Deutsche
Nationalbibliothek.
The Deutsche Nationalbibliothek lists this publica-
tion in the Deutsche Nationalbibliografie;
detailed bibliographic data are available online at
http://dnb.d-nb.de.

None of the content in this book was published in
exchange for payment by commercial parties or
designers; Gestalten selected all included work
based solely on its artistic merit.

This book was printed on paper certified by the FSC®.

Gestalten is a climate-neutral company. We collaborate
with the non-profit carbon offset provider myclimate
(www.myclimate.org) to neutralize the company's carbon footprint produced through our worldwide busi-
ness activities by investing in projects that reduce CO_2
emissions (www.gestalten.com/myclimate).